Population Genetics

Edited by Rafael Trindade Maia and Magnólia Campos de Araújo

Published in London, United Kingdom

Population Genetics
http://dx.doi.org/10.5772/intechopen.98153
Edited by Rafael Trindade Maia and Magnólia Campos de Araújo

Contributors
Mohammad Mia Mukul, Mohammed Abdulwahid Alsoufi, Ge Changrong, Feroz Din Sheikh, Shahnawaz Ali, C Siva, Rafael Trindade Maia, Magnólia Campos de Araújo

First published in London, United Kingdom, 2022 by IntechOpen
IntechOpen is the global imprint of INTECHOPEN LIMITED, registered in England and Wales, registration number: 11086078, 5 Princes Gate Court, London, SW7 2QJ, United Kingdom

British Library Cataloguing-in-Publication Data
A catalogue record for this book is available from the British Library

Additional hard and PDF copies can be obtained from orders@intechopen.com

Population Genetics
Edited by Rafael Trindade Maia and Magnólia Campos de Araújo
p. cm.
Print ISBN 978-1-80355-333-7
Online ISBN 978-1-80355-334-4
eBook (PDF) ISBN 978-1-80355-335-1

Meet the editors

Dr. Rafael Trindade Maia studied Biological Sciences at the Federal Rural University of Pernambuco, Brazil. He received a master´s degree in Genetics, Conservation, and Evolutionary Biology from the National Institute of Amazonian Research, Brazil, in 2008, and a Ph.D. in Animal Biology from the Federal University of Pernambuco, Brazil, in 2013. He is currently an adjunct professor at the Center for the Sustainable Development for Semiarid (CDSA), Federal University of Campina Grande (UFCG), Brazil. He has experience with population genetics, bioinformatics, molecular docking, and modeling and molecular dynamics of proteins. He works in science and biology education. Dr. Maia also leads the Computational and Theoretical Biology (CTB) and Education in Sciences and Biology (ESB) research groups.

Magnólia A. Campos is a biologist with a master's degree in Agronomy/Plant Breeding from the Federal University of Pelotas, Brazil, and a Ph.D. in Biological Sciences/Molecular Biology from the University of Brasília. She has five years of experience in genomic sciences as a postdoctoral researcher at the Federal University of Lavras/Agronomic Institute (IAC), Brazil. Since 2008, she has been a professor at the Federal University of Campina Grande (UFCG), Brazil. She has experience in plant biotechnology, working mainly on genomics, bioinformatics, tissue culture and plant cells, genetic transformation of plants, study of gene expression during plant-microbe interactions, and expression of heterologous proteins in bacteria.

Contents

Preface

Population genetics is a prominent area in biology, as it is a basis for the comprehension of the evolution of living beings. There are several applications of population genetics ranging from the analysis of gene frequencies of a population to the genetic improvement of plants and animals.

This book provides a comprehensive review of population genetics over four sections.

Section 1, "Population Genetics and Main Driving Forces of Evolution", includes chapters on natural selection, derivation, and selection flow.

Section 2, "Population Genetics and Conservation", includes a chapter on phylogeography, biodiversity, molecular markers, and conservation.

Section 3, "Genetic Diversity and Animal Breeding" includes a chapter on the genetic diversity and evolution of Yunnan chicken breeds in China.

Section 4, "Genetic Diversity and Population Structure of Plant Crops", includes a chapter on morpho-nutraceutical diversity of jute and related fiber crops such as vegetables.

Rafael Trindade Maia
Semi-Arid Sustainable Development Center,
Federal University of Campina Grande,
Sumé, Brazil

Magnólia Campos de Araújo
Federal University of Campina Grande,
Sumé, Brazil

Section 1

Population Genetics and Main Driving Forces of Evolution

Chapter 1

Introductory Chapter: Population Genetics

Rafael Trindade Maia and Magnólia Campos de Araújo

1. Introduction

Population genetics is a science that studies the genetic composition and distribution, as well as its effects, through mathematical formulas and indicators for measuring genetic diversity. It aims to evaluate allele, phenotypic and genotypic frequencies in populations of living beings; allowing to understand the origin and dynamics of genetic variation, making it possible to make predictions about the influences of one or more evolutionary processes on these compositions over generations. In this context, population genetics seeks to find an evolutionary meaning to explain genetic variation in living beings and better understand the evolutionary mechanisms that act on it.

Understanding the genetic diversity of populations has several uses, such as monitoring pathogens and vectors, conservation studies and species management, genetic improvement of plants and animals, genetic counselling, monitoring of hereditary diseases, etc. The mechanisms associated with changes in allelic and genotypic frequencies are 1) Mutation; 2) Natural selection; 3) Migration (with gene flow); 4) Genetic drift [1].

2. Mutation

Mutations are the primary sources of genetic variation, responsible for avoiding genetic homogeneity between populations, as they result in new alleles. By definition, mutations are changes in the DNA sequence, which may result from spontaneous errors in DNA replication during cell division or due to external factors such as exposure to mutagenic chemicals, radiation and viral infections [2].

3. Natural selection

Natural selection, initially proposed by Darwin, advocated that those with characteristics that increased the chance of survival or reproduction of individuals tended to settle in populations. In this context, it can be said that natural selection works favouring the advantageous alleles, genotypes and phenotypes and eliminating the unfavourable ones. There are two types of natural selection: positive or Darwinian selection, which acts on an adaptive mutation by increasing its frequency in the population; and negative or purifying selection, which acts in the opposite direction, reducing the frequency or even eliminating deleterious mutations from populations [3].

4. Migration (with gene flow)

The migration of individuals between different populations with consequent reproductive success results in gene flow, which is the transfer of alleles between populations. The outcome of these gene transfers depends on the difference between allelic frequencies in populations and the number and proportion of migrant individuals. It is a crucial evolutionary mechanism for conservation geneticists since gene flow is essential to minimise the effects of inbreeding and genetic drift on natural populations [4].

5. Genetic drift

Of all the evolutionary mechanisms, genetic drift is the one that most randomly alters the gene and allelic frequencies of populations. As it is a completely stochastic process, it is impossible, at first, to predict the allelic frequencies in the face of a drift event. This mechanism is associated with the loss of genetic variation in populations, which may make them more vulnerable in subsequent generations. Genetic drift is a consequence of a drastic alteration of natural and casual order, such as earthquakes, tsunamis, tornadoes, floods, fires, avalanches and other processes, affecting a large population contingent [5].

6. Hardy–Weinberg equilibrium

The Hardy–Weinberg equilibrium is one of the fundamental principles of population genetics. Assuming that a population is panmictic and mated at random and that there is no interference from evolutionary mechanisms (Mutation, Natural Selection, Migration and Drift), allelic and genotypic frequencies remain the same across generations. The Hardy–Weinberg theorem can be applied mathematically to a pair of alleles (a gene) through Newton's binomial, according to the following expression in Eq. (1):

$$p^2 + 2pq + q^2 = 1 \qquad (1)$$

Where:
p is the frequency of the A allele.
q is the frequency of the a allele.
$p2$ is a frequency of AA homozygotes.
q^2 is a frequency of aa homozygotes.
$2pq$ is the frequency of Aa heterozygotes.

When populations are not in Hardy–Weinberg equilibrium, it means that evolutionary forces are acting and changing their allele and genotypic frequencies [6].

Analysing population genetics, through scientific investigations, is one of the best ways to understand the evolutionary history of living beings, with diverse applications in various sectors (health, agriculture, genetic improvement, biotechnology, etc.). In this sense, population genetics is an integrative and challenging science, with great discoveries and challenges to be achieved and a great and promising future in the post-genomic era.

Author details

Rafael Trindade Maia[1,2]* and Magnólia Campos de Araújo[2]

1 Federal University of Campina Grande, Center for Sustainable Development of the Semiarid Region, Sumé, Paraíba State, Brazil

2 Federal University of Campina Grande, Center of Education and Health, Postgraduate Program in Natural Sciences and Biotechnology, Cuité, Paraíba State, Brazil

*Address all correspondence to: rafael.rafatrin@gmail.com

References

[1] Okazaki A, Yamazaki S, Inoue I, et al. Population genetics: Past, present, and future. Human Genetics. 2021;**140**:231-240. DOI: 10.1007/s00439-020-02208-5

[2] Hershberg R. Mutation—The engine of evolution: Studying mutation and its role in the evolution of bacteria. Cold Spring Harbor Perspectives in Biology. 2015;**7**(9):a018077

[3] Gregory TR. Understanding natural selection: Essential concepts and common misconceptions. Evolution: Education and Outreach. 2009;**2**:156-175

[4] Ellstrand NC, Rieseberg LH. When gene flow really matters: Gene flow in applied evolutionary biology. Evolutionary Applications. 2016;**9**(7): 833-836

[5] Lynch M, Ackerman M, Gout JF, et al. Genetic drift, selection and the evolution of the mutation rate. Nature Review Genetics. 2016;**17**:704-714. DOI: 10.1038/nrg.2016.104

[6] Salanti G, Amountza G, Ntzani E, et al. Hardy–Weinberg equilibrium in genetic association studies: An empirical evaluation of reporting, deviations, and power. European Journal of Human Genetics. 2005;**13**:840-848

Section 2

Population Genetics and Conservation

Chapter 2

Perspective Chapter: Molecular Approach for the Study of Genetic Diversity and Conservation Prioritization of Fish Population

Shahnawaz Ali and Chinnathangam Siva

Abstract

Fishes are the most abundant vertebrates in the animal kingdom. They play important biological and ecological roles. Various anthropogenic and climatic factors have led to the decline of natural population and increased the risk of extinction. CBD emphasized the conservation of biodiversity at all levels from genes to ecosystems. However, little attention paid to genetic considerations in restoration efforts. Estimation of genetic diversity and population structure is inevitable for effective implementation of conservation management program. Available DNA markers like mitochondrial and microsatellite markers provide useful insight into understanding the genetic diversity status of fish population in the wild. The present chapter highlights the importance of genetic diversity and its determinants. Utility of mitochondrial and microsatellite markers shown through a case study of a threatened cyprinid species *Neolissochilus hexagonolepis* known as chocolate mahseer that is widely distributed in the North-eastern part of India. Presence of low genetic diversity confirmed its threatened status and further analysis based on various genetic parameters revealed the status of different stocks as well as the population structure of this species. The results obtained could be helpful in rehabilitation and conservation planning and prioritization for the maintenance of a viable population of this species.

Keywords: biodiversity, conservation, population genetics, molecular markers, phylogeography

1. Introduction

Fishes are cold-blooded aquatic vertebrates, and over half of the living vertebrates are fishes. The estimated total number of fishes is more than 35,934 species, and it is higher than the combined total of other vertebrates [1]. Among all known fish species, more than 15,000 are found in freshwater, which is less than 0.3% of available global water, while more than 16,000 species are marine, which is 70% of the earth's surface [2]. The incredible diversity of fishes is astounding that is evident from their morphology, the habitat they occupy, their physiological adaptation, and behavior [3]. They have a long evolutionary history of origin and diversification that began in the Cambrian Period at least 520 million years ago [4]. They occupy all types of the aquatic environment, and to survive and colonize in different habitats, fishes have developed various types of anatomical, physiological, behavioral, and ecological adaptation and plays different types of biological and ecological roles [5]. Fishes are also critically important for the food security and nutrition of ever-increasing human population, and more than 4.5 billion people get at least 15% of their average per capita intake of animal protein from fish. They have unique nutritional properties and are one of the efficient feed converters into high quality food therefore, widely exploited from natural water bodies as well as produced through aquaculture production systems [6].

However, due to various anthropogenic and climatic factors, fish stocks both in marine as well as freshwater are continuously declining and are under severe threats. The unfulfilled demand for resources has resulted in cumulative pressure on the marine ecosystem from a range of human activities. As a result, both marine species and habitat are experiencing detrimental impacts due to different human interference [7–8]. It is estimated that humans have impacted almost 90% of the global ocean surface [9], and marine fish abundances have declined by 38% in last three decades [10]. Loss of coral habitats, overfishing, dredging activities, and damage caused by bottom trawling all have led to the significant decline of marine fish stocks, its recruitment, and yields, and even after continuous efforts recovery has not yet been achieved [11].

On the other side, freshwater ecosystem has been assessed as the most impacted and endangered ecosystem on the planet [12]. The decline in biodiversity is much greater in freshwater than in terrestrial and marine ecosystems [13, 14]. The major threats to global freshwater biodiversity include overexploitation; water pollution; flow modification; destruction or degradation of habitat; invasion by exotic species; infectious diseases, and the combined and interacting influences of these threats have resulted in population declines and range reduction of freshwater biodiversity worldwide [12, 15, 16]. Therefore, direct and indirect anthropogenic impacts have resulted in global decline of biodiversity [7, 17].

In addition, global climate change also poses many threats to biodiversity and alters the physical, chemical, and biological characteristics of freshwater and marine biodiversity and habitats [18, 19]. According to recent estimates, around 50% of global freshwater fish species are potentially threatened due to climate change [20]. The effects of climate change have been recorded and predicted in terms of changes in species phenology, range, and physiology [21], thus accelerating the risk of extinction [22–24]. It is increasingly recognized that the scales of different anthropogenic impacts are greater than natural drivers for the earth system and hence coined the name of a new geological epoch, the "Anthropocene" where human-induced changes dominate over natural cycles [25, 26]. The loss of global biodiversity is now comparable to previous global-scale mass extinction events, and we are now witnessing the "sixth mass extinction" event [27, 28]. It is further shown that the sixth mass extinction is more severe than perceived in terms of species extinction [29].

In its assessment, the Convention on Biological Diversity (CBD) has emphasized the conservation of biodiversity at all levels from genes, population, species, and ecosystems [30, 31]. As discussed above, the threat to freshwater biodiversity is far greater than other ecosystems, and the fish species are becoming either vulnerable or endangered, and their numbers are continuously increasing every year (**Figure 1**), and around 37% of freshwater fishes are threatened with the risk of extinction [20]. The loss of habitat and invasion of non-native species in major riverine systems of

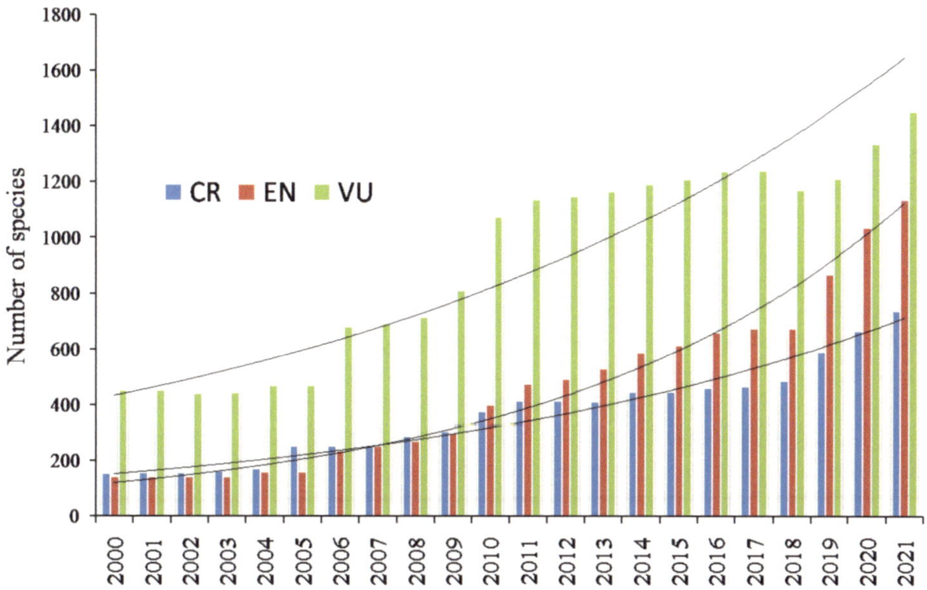

Figure 1.
Changes in numbers of species in the threatened categories (CR: Critical; EN: Endangered; VU: Vulnerable) from 2000 to 2021 (IUCN Red List version 2021–3) for the fish species on the Red List (Accessed on November 17, 2021; https://www.iucnredlist.org/).

the world has reduced not only the species diversity but also caused similarity among species assemblages. This has led to the "taxonomic homogenization" on a regional and global scale [32]. The overall aim of conservation is to protect biological diversity and the underlying processes that sustain it in the face of perturbation caused by human activities. The strategies for conservation need prioritization that maximizes both representation and persistence of diversity [33]. It is known that all organisms are endowed with a genetic blueprint and thus contribute to genetic diversity, which is the foundation of all biological diversity. Earlier works demonstrated that loss of genetic diversity might lead to the collapse of population and even species that are present in the wild [34, 35]. While comparing threatened and non-threatened taxa, it was revealed that the genetic factors such as heterozygosities reduced to a considerable level in the threatened taxa before a species driven to the risk of extinction [36]. Although CBD agreed on the conservation of genetic diversity, little attention has been paid to genetic considerations in restoration efforts, and it remained largely neglected [37]. Therefore, it is necessary to document genetic diversity at the population and species level so that a comprehensive conservation strategy can be implemented for the rehabilitation and restoration of species.

2. Determinants of genetic diversity and its measurement

The vast and varied population of fishes inherit different genetic traits and thus shows remarkable genetic diversity both at spatial and temporal scales [38, 39]. The genetic composition defines the form and functionality of the organism. The presence of genetic variation in the population and species contributes to the ability to respond to environmental changes [40]. The loss of species and their distributional range are detrimental to the genetic diversity, which the species inherited and accumulated over millions of years of evolutionary processes. Thus documenting the genetic variation in populations is important to understand the forces that change their genetic composition over time, and thus their evolutionary relationship is described through the study of population genetics [41]. It is also important to understand that each individual of a species might have a similar phenotype but distinct genetic makeup. These differences arise due to the difference in their nucleotide sequences (e.g. DNA sequences) which is called "polymorphism." Genetic diversity provides the raw material for the survival, evolution, and natural selection of the organism [42]. One of the important phenomenons, which contribute towards the genetic variation among individuals at species or population level, is "mutation." In other words, it is a base pair substitution in the DNA sequences (either coding or non-coding region) during the replication, and this is an essential requirement for the evolution [41]. Mutation in non-coding sequences evolves faster than coding sequences since it does not directly affect the gene functions. Thus these genetic variants or "alleles" appear or added at each generation due to random mutation or may disappear due to loss of alleles under the influence of "genetic drift" (i.e., a random change in gene frequencies). The other important evolutionary force responsible for high genetic variation is the "Natural selection" that can change gene frequencies in the population and leads toward the relative fitness in the population. However, for natural selection to affect the allele frequency, the locus must be in the coding region [43]. In contrast to this, "Neutral Theory of Molecular Evolution" proposed by "Kimura" [44] argues that most allelic variations and substitutions in proteins and DNA are neutral. According to this theory, gene frequencies may change by "genetic drift" without the influence of natural selection, and in a large panmictic population (i.e., where species show random mating within a population), and it is inversely proportional to the effective population size [45].

The detection and measurements of genetic diversity and population structure are essentially required for the development of the appropriate strategies for the implementation of conservation programs [46, 47]. Furthermore, molecular phylogenetics and genetic diversity analysis help in ascertaining the taxonomic identity and evolutionary relationship of the wild species. There are certain population genetic parameters that are measured for the evaluation of genetic variability at individual and population levels. These measures are essential for the comprehensive assessment of genetic structure within and among the population. Among such important population genetic parameters are the percentage of polymorphic loci; the number of alleles per locus; the effective number of alleles per locus; observed and expected heterozygosity; estimates of effective population size; and assessment of linkage disequilibrium. Further, for estimating variations between population it is essentially required to measure different types of variances (Fst, G_{ST}, R_{ST}), genetic distances, and correlation between genetic distance and geographic distance [43, 48]. These population genetic parameters provide important data to draw any plausible conclusion about the status of the stock and its genetic structure.

3. Application of molecular technology and fisheries genetics

The advent of molecular techniques in the last seven decades has provided significant insights into the population structure and its genetic diversity. Initially, the technique of protein gel electrophoresis to several allozyme loci was applied to measure the genetic variation in the species [49] and assessment of the fish genetic stock. The use of allozyme remained a dominant method until the development of

DNA amplification using the PCR (Polymerase Chain Reaction) technique [50]. The arrival of PCR-based techniques revolutionized the field of molecular genetics and led to the emergence of fish genomics by the use of DNA-based markers technology. A genetic or molecular marker is a gene or DNA sequence with a known location on a chromosome and associated with a particular gene or trait. The popular genetic markers widely used for genetic diversity assessment include allozymes, mitochondrial DNA, RFLP, RAPD, AFLP, microsatellite, SNP, and EST markers. Genetic markers have been applied to three areas of fisheries in particular; stock structure analysis, aquaculture, and taxonomy/systematics [51] with varying degrees of success [52].

Among the different available DNA markers, mitochondrial DNA (mtDNA) has been widely and effectively used for the assessment of population structure and phylogenetic study [53, 54]. Mitochondrial DNA (mtDNA), as the name suggests, is contained in the mitochondria of the cell and is generally maternally inherited. The general features of mitochondrial DNA include predominantly female inheritance, lack of recombination, selectively neutral, high rate of evolution, relatively simple structure, and multiple copies in the cell [55, 56]. Therefore, different mtDNA gene sequences have proved to be a good marker for analyzing variation at interspecific and intraspecific levels in fishes. Mitochondrial DNA marker (mtDNA) is widely used to study the gene flow, hybrid zones, population structure, phylogenetics, phylogeography, molecular evolution, and conservation genetics [57]. Another type of marker, which is known as satellite DNA is increasingly used for the investigation of genetic variability and divergence between the species [58]. Microsatellite, also known as Simple Sequence Repeats (SSRs), has widely been utilized for studies in population genetics, evolutionary and conservation biology of species and therefore considered as the most significant genetic marker [59]. A microsatellite is tandem repeated motifs of 1–6 bases found in all prokaryotic and eukaryotic genomes. They are present in both coding and non-coding regions and are usually characterized by a high degree of length polymorphism. Further, microsatellites featured with co-dominant inheritance, inheritance in a mendelian fashion, wide distribution, high stability, and repeatability signify their usage for the assessment of genetic diversity within and between populations and provide significant genetic information [60]. Hence, species-specific microsatellite markers are extensively developed and studied in different fish populations [61, 62].

Recently, with the advent of next-generation sequencing (NGS) platforms, the SSR markers in the non-model organism can be developed rapidly and efficiently compared to the conventional methods [63, 64]. The random sequencing based approach also facilitates the genotyping of a high number of loci at moderate costs [65]. Among the different NGS platforms available, the Illumina sequencing method is a powerful tool for the discovery of SSRs and delivers the highest yield of error-free data for the most sensitive or complex sequencing samples [66].

4. Conservation prioritization of fish population: a case study

The use of molecular markers based technology has immensely contributed to different aspects of conservation genetics of species, such as resolving the taxonomic ambiguity; designing captive and marker assisted breeding programs; detecting diversity within and among geographical populations; estimating gene flow, and understanding the factors contributing to fitness [67]. Therefore, management of the species must include information on the extent and organization of genetic diversity in populations to suggest sustainable conservation strategies. This becomes more relevant when we are dealing with endangered species [46]. The fundamental aim of the conservation of species is to minimize genetic deterioration of endangered stock and maintain a viable population to avoid the bottleneck and risk of extinction. The parameters such as genetic divergence among populations and gene flow rate are helpful in characterizing populations, species, and subspecies in different conservation units [68]. Here we have briefly discussed the use of mitochondrial and microsatellite markers in conservation prioritization of a threatened cyprinid species *Neolissochilus hexagonolepis* known as chocolate mahseer that is widely distributed in the north-eastern part of India.

Mahseer is the common name used for three carp genera, *viz. Tor*, *Neolissochilus*, and *Naziritor* (family Cyprinidae). *N. hexagonolepis* has been widely reported from southeast Asia and is abundantly available in the Brahmaputra river basin of Northeast India [69]. The species is enjoyed as food as well as sports fish and also identified as a candidate species for aquaculture [70]. However, the natural population is rapidly declining due to various anthropogenic reasons such as degradation of natural habitat, hydro development projects, and angling demands of the species and therefore categorized as threatened species by IUCN [71, 72]. For the implementation of any effective conservation program it is inevitable to obtain basic genetic information of this species. Therefore, we used mitochondrial and microsatellite DNA markers to study the genetic structure, population history, and genetic diversity of geographically isolated populations of the *N. hexagonolepis* from Northeast India. The information provided to identify genetically diverse stocks as well as delineation of conservation units that can be utilized to optimize the conservation of the chocolate mahseer.

In the experimental setup, 200 fish samples were collected from different geographically isolated drainages from Northeast India. First, we evaluated the targeted genetic parameters using three mitochondrial markers, namely ATPase6/8, cytochrome oxidase I (CO-I) and cytochrome b (Cytb). For amplification of these mitochondrial genes, total genomic DNA from fin samples was used and amplified using standard primer pairs designed from the whole mitochondrial genome sequence. Further, PCR products were column purified and sequenced in both directions using an ABI 3130 Genetic Analyzer (Applied Biosystem, Carlsbad, CA) with Big Dye Terminator cycle sequencing kit v.3.1 with the help of the same primers used for amplification of the target genes. Robust statistical analysis was performed using suitable computer programs to estimate the population genetic parameter, mainly polymorphic sites (S), haplotype diversity (H_d), nucleotide diversity (p), and haplotype number. Molecular variance (AMOVA), molecular diversity indices, and genetic differentiation (F_{ST}) were also calculated. Moreover, the phylogenetic relationship among individuals of different populations was constructed by implementing the maximum likelihood tree method (MLM) based on the best-predicted model [73]. Mean genetic distances between the populations for all the three genes were also calculated. Geographical distances were simulated with haplotypes to determine the optimal number of population groups (K = 2–8). Possible correlation between pairwise genetic differentiation and geographical distances among nine populations was estimated by applying the Mantel test [74].

The analysis revealed the genetic diversity status of different populations based on their haplotype and nucleotide diversity pattern. In comparison, some populations has undergone a reduction in size or recent colonization events whereas, other

populations showed a high level of divergence between haplotypes, indicating a long historical evolutionary pattern [75]. Analysis of molecular variance revealed a high level of genetic structuring among populations. Five major groups and one paraphyletic intermediate group were obtained by phylogenetic analysis. Overall results indicated a positive correlation between geographical distances and genetic divergence [76]. The analysis of different genetic parameters clearly indicated that most of the variation in genetic differentiation is present among population groups, and genetic variations in the chocolate mahseer population might be due to specific habitat conditions that influence population genetic structure. Further, the study also confirmed the threatened status of the population being low in genetic diversity. Thus, information generated by the study would be helpful for developing stock-specific strategies for
N. hexagonolepis breeding, conservation, and management.

Further, microsatellite markers were developed, and population genetics, evolutionary, and conservation biology of *N. hexagonolepis* were studied. Here we used NGS technology (Illumina Miseq) to develop 25 novel SSR markers and further used these markers for assessing the genetic diversity and population structure of this species from Northeast India. Different population genetic parameters such as a number of polymorphic loci, numbers of alleles per locus, observed and expected heterozygosity, and pairwise genetic diversity were estimated using available computer programs. Population structure and bottleneck were estimated as well as migration pattern was studied using appropriate statistical methods.

In *N. hexagonolepis* we found tetra-nucleotides as the most frequent microsatellite motifs that were opposed to what is observed for other cyprinids where di-nucleotide is the most abundant [77]. All the loci were highly polymorphic and thus used for the analysis. In certain population, observed heterozygosity was lower than expected heterozygosity (He) which indicated the inbreeding effect within the populations [78]. Analysis of molecular variance (AMOVA) and high *F*st indicated relatively low gene flow among the population. Migration analysis also revealed that there is no active migration among the studied populations of *N. hexagonolepis*. The STRUCTURE analysis identified five subgroups that substantiate the result of cluster analysis and factorial correspondence analysis (FCA). Based on the estimation of different genetic parameters through statistical analysis, we could successfully identify the genetic status of different stocks as well as the population structure of this species [79]. The identified major groups can be considered as different conservation units when applying any management measures. Thus both the markers were extremely useful in genetic stock assessment of the species under study and provided critical information regarding conservation prioritization.

5. Conclusion

Although freshwater ecosystems is under severe threat due to various anthropogenic and climatic factors, little attention and effort have been paid towards its conservation. Identification of fish stock structure and assessment of their genetic diversity should be an important component of any conservation planning. Available molecular tools are quite useful for the estimation of species diversity at individual to population levels. Further, it becomes more important when we deal with species of endangered categories. The present study clearly indicated that the use of mitochondrial and microsatellite markers has provided a great deal of information related to the fish population under study. We could develop novel microsatellite markers, which will be further useful for stock characterization as well as any marker-assisted breeding program in aquaculture. Apart from these, molecular techniques are commonly used as bio-monitoring tools for assessing the genetic diversity status of the species that help in rehabilitation and conservation planning and prioritization for the maintenance of a sustainable ecosystem.

Acknowledgements

The authors acknowledge the financial assistance provided by Indian Council of Agricultural Research (ICAR) New Delhi, India for carrying out this research work under the project "Fish Genetic Stock—An outreach activity". The authors express thanks to Director, ICAR-Directorate of Coldwater Fisheries Research, Bhimtal, India for providing all necessary support.

Conflict of interest

The authors declare no conflict of interest.

Author details

Shahnawaz Ali* and Chinnathangam Siva
ICAR-Directorate of Coldwater Fisheries Research, Bhimtal, Uttarakhand, India

*Address all correspondence to: alicife@gmail.com

References

[1] IUCN. Red List (version 2021-3) for the Fish Species on the Red List. 2021. Available from: https://www.iucnredlist.org/ [Accessed: November 17, 2021]

[2] Levêque C, Oberdorff T, PAUGy D, MLJ S, Tedesco PA. Global diversity of fish (Pisces) in freshwater. Hydrobiologia. 2008;**595**:545-567. DOI: 10.1007/s10750-007-9034-0

[3] Nelson JS, Grande TC, Wilson MV. Fishes of the World. Hoboken, New Jersey: John Wiley & Sons; 2016

[4] Long J, Choo B, Clement A. The evolution of fishes through geological time. In: Johanson Z, Underwood C, Richter M, editors. Evolution and Development of Fishes. Cambridge: Cambridge University Press; 2019. pp. 3-29. DOI: 10.1017/9781316832172.002

[5] Helfman G, Collette BB, Facey DE, Bowen BW. The Diversity of Fishes: Biology, Evolution, and Ecology. Hoboken, NJ: John Wiley & Sons; 2019. p. 693

[6] Béné C, Barange M, Subasinghe R, Pinstrup-Andersen P, Merino G, Hemre GI, et al. Feeding 9 billion by 2050–Putting fish back on the menu. Food Security. 2015;**7**(2):261-274. DOI: 10.1007/s12571-015-0427-z

[7] Butchart SH, Walpole M, Collen B, Van Strien A, Scharlemann JP, Almond RE, et al. Global biodiversity: Indicators of recent declines. Science. 2010;**328**(5982):1164-1168. DOI: 10.1126/science.1187512

[8] Halpern BS, Frazier M, Potapenko J, Casey KS, Koenig K, Longo C, et al. Spatial and temporal changes in cumulative human impacts on the world's ocean. Nature Communication. 2015;**6**(1):1-7. DOI: 10.1038/ncomms8615

[9] Jones KR, Klein CJ, Halpern BS, Venter O, Grantham H, Kuempel CD, et al. The location and protection status of Earth's diminishing marine wilderness. Current Biology. 2018;**28**(15):2506-2512. DOI: 10.1016/j.cub.2018.06.010

[10] Hutchings JA, Minto C, Ricard D, et al. Trends in the abundance of marine fishes. Canadian Journal of Fisheries and Aquatic Science. 2010;**67**:1205-1210. DOI: 10.1139/F10-081

[11] Britten GL, Duarte CM, Worm B. Recovery of assessed global fish stocks remains uncertain. Proceedings of the National Academy of Sciences of the United States of America. 2021;**118**(31):e2108532118. DOI: 10.1073/pnas.2108532118

[12] Dudgeon D, Arthington AH, Gessner MO, Kawabata ZI, Knowler DJ, Leveque C, et al. Freshwater biodiversity: Importance, threats, status and conservation challenges. Biological Review. 2006;**81**:163-182. DOI: 10.1017/S1464793105006950

[13] Sala OE, Chapin FS, Armesto JJ, Berlow R, Bloomfield J, Dirzo R, et al. Global biodiversity scenarios for the year 2100. Science. 2000;**287**:1770-1774

[14] Malmqvist B, Rundle S. Threats to the running water ecosystems of the world. Environmental Conservation. 2002;**29**(2):134-153. DOI: 10.1017/S0376892902000097

[15] Reid AJ, Carlson AK, Creed IF, Eliason EJ, Gell PA, Johnson PT, et al. Emerging threats and persistent

conservation challenges for freshwater biodiversity. Biological Reviews. 2019;**94**(3):849-873. DOI: 10.1111/brv.12480

[16] Barbarossa V, Schmitt RJ, Huijbregts MA, Zarfl C, King H, Schipper AM. Impacts of current and future large dams on the geographic range connectivity of freshwater fish worldwide. Proceedings of the National Academy of Sciences of the United States of America. 2020;**117**(7):3648-3655. DOI: 10.1073/pnas.1912776117

[17] Su G, Logez M, Xu J, Tao S, Villéger S, Brosse S. Human impacts on global freshwater fish biodiversity. Science. 2021;**371**(6531):835-838. DOI: 10.1126/science.abd3369

[18] Oberdorff T, Jézéquel C, Campero M, Carvajal-Vallejos F, Cornu JF, Dias MS, et al. Opinion Paper: How vulnerable are Amazonian freshwater fishes to ongoing climate change? Journal of Applied Ichthyology. 2015;**31**:4-9

[19] Freer JJ, Partridge JC, Tarling GA, Collins MA, Genner MJ. Predicting ecological responses in a changing ocean: The effects of future climate uncertainty. Marine Biology. 2018;**165**(1):1-18

[20] Darwall WR, Freyhof J. Lost fishes, who is counting? The extent of the threat to freshwater fish biodiversity. In: Closs GP, Krkosek M, Olden JD, editors. Conservation of Freshwater Fishes. Cambridge: Cambridge University Press; 2015. pp. 1-35

[21] Bellard C, Bertelsmeier C, Leadley P, Thuiller W, Courchamp F. Impacts of climate change on the future of biodiversity. Ecology Letters. 2012;**15**(4):365-377. DOI: 10.1111/j.1461-0248.2011.01736.x

[22] Parmesan C, Yohe G. A globally coherent fingerprint of climate change impacts across natural systems. Nature. 2003;**421**(6918):37-42. DOI: 10.1038/nature01286

[23] Urban MC. Accelerating extinction risk from climate change. Science. 2015;**348**(6234):571-573

[24] Wiens JJ. Climate-related local extinctions are already widespread among plant and animal species. PLoS Biology. 2016;**14**(12):e2001104. DOI: 10.1371/journal.pbio.2001104

[25] Crutzen PJ. Geology of mankind. Nature. 2002;**415**:23

[26] Luypaert T, Hagan JG, McCarthy ML, Poti M. Status of marine biodiversity in the Anthropocene. In: YOUMARES 9-The Oceans: Our Research, Our Future. Cham: Springer; 2020. pp. 57-82

[27] Barnosky AD, Matzke N, Tomiya S, et al. Has the Earth's sixth mass extinction already arrived? Nature. 2011;**471**:51-57. DOI: 10.1038/nature09678

[28] Ceballos G, Ehrlich PR, Barnosky AD, et al. Accelerated modern human-induced species losses: Entering the sixth mass extinction. Sciences Advances. 2015;**1**:e1400253. DOI: 10.1126/sciadv.1400253

[29] Ceballos G, Ehrlich PR, Dirzo R. Biological annihilation via the ongoing sixth mass extinction signaled by vertebrate population losses and declines. Proceedings of the National Academy of Sciences of the United States of America. 2017;**114**(30):E6089-E6096. DOI: 10.1073/pnas.1704949114

[30] Convention on Biological Diversity. Framework for monitoring implementation of the achievement of the 2010 target and integration of targets into the thematic programmes of work,

COP 8 Decision VIII/15. 2006. Available from: www.cbd.int/decisions

[31] Walpole M, Almond RE, Besançon C, Butchart SH, Campbell-Lendrum D, Carr GM, et al. Tracking progress toward the 2010 biodiversity target and beyond. Science. 2009;**325**(5947):1503-1504

[32] Villéger S, Blanchet S, Beauchard O, Oberdorff T, Brosse S. Homogenization patterns of the world's freshwater fish faunas. Proceedings of the National Academy of Sciences of the United States of America. 2011;**108**(44):18003-18008

[33] Moritz C. Strategies to protect biological diversity and the evolutionary processes that sustain it. Systematic Biology. 2002;**51**(2):238-254. DOI: 10.1080/10635150252899752

[34] Briskie JV, Mackintosh M. Hatching failure increases with severity of population bottlenecks in birds. Proceedings of the National Academy of Science of the United States of America. 2004;**101**:558-561

[35] Frankham R. Genetics and extinction. Biological Conservation. 2005;**126**:131-140

[36] Spielman D, Brook BW, Frankham R. Most species are not driven to extinction before genetic factors impact them. Proceedings of the National Academy of Sciences of the United States of America. 2004;**101**(42):15261-15264

[37] Laikre L, Allendorf FW, Aroner LC, Baker CS, Gregovich DP, et al. Neglect of genetic diversity in implementation of the convention on biological diversity. Conservation Biology. 2010;**24**(1):86-88

[38] Riccioni G, Landi M, Ferrara G, Milano I, Cariani A, Zane L, et al. Spatio-temporal population structuring and genetic diversity retention in depleted Atlantic bluefin tuna of the Mediterranean Sea. Proceedings of the National Academy of Sciences of the United States of America. 2010;**107**(5):2102-2107. DOI: 10.1073/pnas.0908281107

[39] Palstra FP, Ruzzante DE. A temporal perspective on population structure and gene flow in Atlantic salmon (*Salmo salar*) in Newfoundland, Canada. Canadian Journal of Fisheries and Aquatic Sciences. 2010;**67**(2):225-242

[40] Hughes AR, Inouye BD, Johnson MT, Underwood N, Vellend M. Ecological consequences of genetic diversity. Ecology Letters. 2008;**11**(6):609-623. DOI: 10.1111/j.1461-0248.2008.01179.x

[41] Burden CJ. Population genetics. In: Ranganathan S, Gribskov M, Nakai K, Schönbach C, editors. Encyclopedia of Bioinformatics and Computational Biology. Vol. 2. Amsterdam: Elsevier; 2019. pp. 759-788. DOI: 10.1016/B978-0-12-809633-8.90693-9

[42] Fisher RA. The Genetical Theory of Natural Selection. Oxford, United Kingdom: Clarendon Press; 1930. DOI: 10.5962/bhl.title.27468

[43] Gillespie JH. Population Genetics: A Concise Guide. Baltimore and London: The Johns Hopkins University Press; 2004

[44] Kimura M. The Neutral Theory of Molecular Evolution. Cambridge: Cambridge University Press; 1983

[45] Ellegren H, Galtier N. Determinants of genetic diversity. Nature Reviews Genetics. 2016;**17**(7):422-433

[46] Meffe GK. Conservation genetics and the management of endangered fishes. Fisheries. 1986;**11**(1):14-23

[47] Milligan BG, Leebens-Mack J, Strand AE. Conservation genetics:

Beyond the maintenance of marker diversity. Molecular Ecology. 1994; **3**:423-435

[48] Höglund J. Evolutionary Conservation Genetics. Oxford: Oxford University Press; 2009

[49] Lewontin RC, Hubby JL. A molecular approach to the study of genetic heterozygosity in natural populations. II. Amount of variation and degree of heterozygosity in natural populations of *Drosophila pseudoobscura*. Genetics. 1966;**54**:595-609

[50] Ferguson A, Taggart JB, Prodohl PA, McMeel O, Thompson C, Stone C, et al. The application of molecular markers to the study and conservation of fish populations, with special reference to Salmo. Journal of Fish Biology. 1995;**47**:103-126

[51] Ward RD, Grewe M. Appraisal of molecular genetic techniques in fisheries. In: Carvalho GR, Pitcher TJ, editors. Molecular Genetics in Fisheries. London: Chapman & hall; 1994. pp. 29-54

[52] Liu ZJ, Cordes JF. DNA marker technologies and their applications in aquaculture genetics. Aquaculture. 2004;**238**(1-4):1-37

[53] Barat A, Ali S, Sati J, Sivaraman GK. Phylogenetic analysis of fishes of the subfamily Schizothoracinae (Teleostei: Cyprinidae) from Indian Himalayas using *cytochrome b* gene. Indian Journal of Fish. 2012;**59**:43-47

[54] Zheng LP, Yang JX. Genetic diversity and population structure of *Acrossocheilus yunnanensis* (Teleostei, Cyprinidae) inferred from four mitochondrial gene sequences. Mitochondrial DNA Part A. 2018;**29**(4):606-614

[55] Avise JC. Molecular Markers, Natural History and Evolution. New York: Chapman & Hall; 1994. DOI: 10.1007/978-1-4615-2381-9

[56] Birky CW, Fuerst P, Maruyama T. Organelle gene diversity under migration, mutation, and drift equilibrium expectations, approach to equilibrium, effects of heteroplasmic cells, and comparison to nuclear genes. Genetics. 1989;**121**:613-627

[57] Rosetti N, Remis MI. Spatial genetic structure and mitochondrial DNA phylogeography of Argentinean populations of the grasshopper Dichroplus elongatus. PLoS ONE. 2012;**7**(7):e40807

[58] Tautz D. Hypervariability of simple sequences as a general source for polymorphic DNA markers. Nucleic Acids Research. 1989;**17**:6463-6471

[59] Baillie SM, Muir AM, Scribner K, Bentzen P, Krueger CC. Loss of genetic diversity and reduction of genetic distance among lake trout *Salvelinus namaycush ecomorphs*, Lake Superior 1959 to 2013. Journal of Great Lakes Research. 2016;**42**(2):204-216

[60] Chistiakov DA, Hellemans B, Volckaert FAM. Microsatellites and their genomic distribution, evolution, function and applications: A review with special reference to fish genetics. Aquaculture. 2006;**255**:1-29

[61] Zheng X, Kuang Y, Lü W. Transcriptome-derived EST–SSR markers and their correlations with growth traits in crucian carp *Carassius auratus*. Fisheries Science. 2014;**80**(5):977-984

[62] Ariede RB, Freitas MV, Hata ME, Matrochirico-Filho VA, Utsunomia R, Mendonça FF, et al. Development of microsatellite markers using next-generation sequencing for the fish

Colossoma macropomum. Molecular Biology Reports. 2018;**45**(1):9-18

[63] Ekblom R, Galindo J. Applications of next generation sequencing in molecular ecology of non-model organisms. Heredity. 2011;**107**:1-15. DOI: 10.1038/hdy.2010.152

[64] Heras S, Planella L, Caldarazzo I, Vera M, García-Marín J, Roldán MI. Development and characterization of novel microsatellite markers by Next Generation Sequencing for the blue and red shrimp *Aristeus antennatus*. Peer J. 2016;**4**:e2200. DOI: 10.7717/peerj.2200

[65] Guichoux E, Lagache L, Wagner S, Chaumeil P, Léger P, Lepais O, et al. Current trends in microsatellite genotyping. Molecular Ecology Resources. 2011;**11**:591-611. DOI: 10.1111/j.1755-0998.2011.03014.x

[66] Ergüner B, Ustek D, Sagiroglu MS. Performance comparison of Next Generation sequencing platforms. In: 2015 37th Annual International Conference of the IEEE Engineering in Medicine and Biology Society. Milan, Italy: IEEE; 2015. pp. 6453-6456. DOI: 10.1109/EMBC.2015.7319870

[67] Vrijenhoek RC. Conservation genetics of freshwater fish. Journal of Fish Biology. 1998;**1998**(53):394-412. DOI: 10.1111/j.1095-8649.1998.tb01039.x

[68] Moritz C. Defining 'Evolutionarily Significant Units' for conservation. Trends in Ecology and Evolution. 1994;**9**:373-375

[69] Laskar BA, Bhattacharjee MJ, Dhar B, Mahadani P, Kundu S, Ghosh SK. The species dilemma of northeast Indian mahseer (Actinopterygii: Cyprinidae): DNA barcoding in clarifying the riddle. PloS One. 2013;**8**(1):e53704. DOI: 10.1371/journal.pone.0053704

[70] Sarma D, Akhtar MS, Das P, Ganesh G, Ciji A, Mahanta PC. Observations on larval development of chocolate mahseer (*Neolissochilus hexagonolepis*) (McClelland, 1839). Indian Journal of Fisheries. 2015;**62**(1):135-138

[71] Mahapatra BK, Vinod K, Mandal BK. Studies on chocolate mahseer, Neolissocheilus hexagonolepis (McClelland) fishery and the cause of its decline in Umiam reservoir, Meghalaya. Journal of Natcon. 2004;**16**(1):199-205

[72] Abbasi T, Abbasi SA. Small hydro and the environmental implications of its extensive utilization. Renewable and Sustainable Energy Reviews. 2011;**15**(4):2134-2143

[73] Tamura K, Stecher G, Peterson D, Filipski A, Kumar S. MEGA6: Molecular Evolutionary Genetics Analysis Version 6.0. Molecular Biology and Evolution. 2013;**30**:2725-2729

[74] Excoffier L, Laval G, Schneider S. Arlequin (version 3.0): An integrated software package for population genetics data analysis. Evolutionary Bioinformatics. 2005;**1**:47-50

[75] Grant WAS, Bowen BW. Shallow population histories in deep evolutionary lineages of marine fishes: Insights from sardines and anchovies and lessons for conservation. Journal of Heredity. 1998;**89**(5):415-426. DOI: 10.1093/jhered/89.5.415

[76] Sharma L, Ali S, Barat A, Kumar R, Pande V, Laskar MA, et al. Molecular identification and genetic diversity analysis of Chocolate mahseer (*Neolissochilus hexagonolepis*) populations of Northeast India, using mitochondrial DNA markers. Mitochondrial DNA Part A. 2019;**30**(3):397-406. DOI: 10.1080/24701394.2018.1526929

[77] Sahu BP, Sahoo L, Joshi CG, Mohanty P, Sundaray JK, Jayasankar P, et al. Isolation and characterization of polymorphic microsatellite loci in Indian major carp, *Catla catla* using next generation sequencing platform. Biochemical Systematics and Ecology. 2014;**57**:357-362

[78] Barat A, Sahoo PK, Kumar R, Mir JI, Ali S, Patiyal RS, et al. Molecular characterization of rainbow trout, *Oncorhynchus mykiss* (Walbaum, 1792) stocks in India. Journal of Genetics. 2015;**94**:13-18

[79] Sharma L, Ali S, Siva C, Kumar R, Barat A, Sahoo PK, et al. Genetic diversity and population structure of the threatened chocolate mahseer (*Neolissochilus hexagonolepis* McClelland 1839) based on SSR markers: implications for conservation management in Northeast India. Molecular Biology Reports. 2019;**46**(5): 5237-5249. DOI: 10.1007/s11033-019-04981-7

Section 3

Genetic Diversity and Animal Breeding

Chapter 3

Status of World's Unique Animal Genetic Resource of Ladakh

Feroz Din Sheikh

Abstract

Ladakh is the only typical cold arid zone of our country with extreme climate and hostile condition, the area makes its flora and fauna a unique one and distinct from rest of the country. Around 60% of the farmers of Ladakh are Pastoralist and 40% of the farmers are agriculturist and the majority of the economic income comes from animal rearing in Ladakh. It has some of the world's best animal genetic resource in the form of Changthangi Pashmina goats, Changthangi Sheep, Malra Goat, Malluk and Purig Sheep, Semi domesticated Yak and its crosses, Bactrian (Double Humped Camel) Zanskari horse, Ladakhi Cattle and Changthangi Dog. All these livestock contributes a lot to the income of the farmers of Ladakh especially the Changthang nomads who are completely dependent on livestock rearing. The livestock in Changthang is reared on extensive system on high altitude pastureland. During the past few decades these unique germplasms are facing several threats for its eco-friendly existence with the human populations. These threats and constraints are figured with possible recommendation and solution in the present study. Ladakh has been deprived of basic research facilities in animal science sector for so many years due to which this unique genetic resources are declining in terms of numbers as well as in production. Another reason is shifting of Ladakh economy from agro pastoralist to tourism business. If necessary steps are not taken immediately a time will reach that all this precious animal will be lost forever. The present article describes the present status and critical issues pertaining to animal genetic resource of Ladakh.

Keywords: Ladakh, Changthang, animals, population, livestock

1. Introduction

Nature has its own role to play in selecting the best germplasm which can thrive and which are adapted to extreme type environment. The process of selection though it takes thousands of year but the result is a birth of a perfect organism, one such unique example is the highly diverse and unique domestic and wild animals of Ladakh. Ladakh meaning "Land of Passes" covers around 45,000 square miles (117,000 sq.km) and contains the Ladakh range, which is south-eastern extension of the Karakoram Range and upper Himalayan Range. It is administratively divided between Pakistan (northwest), as part of Gilgit Baltistan, and India (southeast), as part of Ladakh Union Territory (until October 31, 2019, part of Jammu Kashmir State; in addition, China administers portion of north-eastern part of Ladakh. Ladakh has

Figure 1.
Map of Union Territory of Ladakh.

two main districts Leh and Kargil and it is the only typical cold arid zone of our country. The total land area of Leh district alone is 45,110 km^2 and along with Kargil district it forms more than 70% area of erstwhile Jammu & Kashmir state (**Figure 1**). The geographical location with an altitude extending from 11,000 to 16,000 ft asl and typical climatic condition with temperature ranging from +40°C in summer to −40°C of the area makes its floura and fauna a unique place on mother earth. Around 60% of the farmers of Ladakh are Pastoralist and 40% of the farmers are agriculturist and the majority of the economic income comes from animal rearing in Ladakh [1]. It has some of the world's best animal genetic resource in the form of Changthangi Pashmina goats, Changthangi Sheep, Malra Goat, Malluk Sheep, Semi domesticated Yak and its crosses, Bactrian (Double Humped Camel) Zanskari horse, Ladakhi Cattle and Changthangi Dog. These livestock are the main source of income for Ladakhi people and in the case of Changthang people, they are completely dependent on livestock rearing. The present status of each of this unique species are presented below:

2. Changthangi Pashmina goat

The Pashmina internationally known as "Cashmere", a fine luxury fibre, is the prince of the specialty fibres obtained from domestic goats known as "*capra ibex*". The word "pashmina" comes from "pashm" meaning wool in Persian language and it is also known as 'tivit' sometimes. The cashmere breeds of goats (**Figure 2**) are found throughout the world and some of the important breeds are, Vatani of Afghanistan, Don, Orenberg and Altai mountain of Russia, Tan goats of China, Markhore and Raini of Iran, Kurdi of Iraq, Feral goats of Australia, Chegu and Changthangi goats of India. The three major producers of cashmere wool are China (60–70%), Mongolia (30–40%) and Iran, Afghanistan, Nepal etc (10–20%). Ladakh is a world leader in producing the

Figure 2.
Changthangi Pashmina Goat.

finest Pashmina (fibre diameter less than 12μ) and a store house of the best germplasm with potential to exploit the superior genetic material for improvement of a whole range of pashmina producing goats in the world [2]. Changthangi goats also called as 'Changra' goats is primarily used for Pashmina production and secondarily used for meat production purpose, they also produce some amount of milk used by the nomads for making butter, cheese and curd for domestic use. These Pashmina goats are of great importance for vitalizing the economy of poverty stricken region of Changthang region of Leh districts of Ladakh (**Table 1**). In our country the Changthang region produces around 45,000 kgs of raw Pashmina fibre every year from about 2.03 lacs of Changthangi goats [3], which forms 99% of the total Pashmina production of India with 1% from Chegu breed of goat from Himachal Pradesh state. These goats are reared by Changpa Nomads on vast pastureland of Changthang region of Ladakh India (**Figure 3**). Compared to rest of the world the Pashmina fibre produced from this goats are relatively longer and finer, which aids the Kashmiri weavers to spin easily for preparing various high quality winter garments. This fibre has three times insulating values of the finest wool on a comparable weight basis and also possess 90% of the strength of merino wool and 60% strength of mohair [4]. The importance of this valuable fibre was discovered by the people of Europe during 17th century. It is believed that emperor Napoleon presented a shawl to Empress Josephine and there is a rumor that pashmina garments are as soft as baby skin and provide warm to hatch an egg. The craftsmen in Kashmir developed their own technique to process the raw material and used it for the manufacture of shawls. Thousands of artisans get involved in Kashmir for preparing shawls and pashmina business sustained the economy of thousands of artisan families in Kashmir and thousands of Changpa families in Ladakh as the possibilities for agriculture farming in Changthang is quite low [5]. The People of Changthang rear huge herds of goats and remained in migration in specific routes in search of better pasture

Livestock Product/ Village	Wool	Pashmina	Mutton	Chevon	Beef	Total
Nyoma	3649.33	11600.00	23460.00	13050.00	2440.00	54199.33
	+910.86	+3639.07	+5855.52	+4093.96	+1116.00	+14790.00
Nidder	4289.60	24336.00	27576.00	27378.00	912.00	84491.60
	+1183.78	+2891.57	+7610.00	+3253.02	+307.35	+1 1006.43
Mud	6205.81	30305.455	39894.54	40936.30	1156.36	111655.81
	+2044.7	+6150.71	+13145.02	+6150.7 1	+364.43	+22760.24
Rongo	3201.33	37300.00	20580.00	41962.50	760.00	103803.83
	+405.80	+9626.53	+2608.76	+10829.84	+362.8 I	+21224.95
Koyul	4786.60	43800.00	30771.00	49275.00	.3570.00	132202.60
	+1407.29	+5752.61	+9046.91	+6471.68	+801.73	+18265.05
Kharnak	3676.4	42876.00	23634.00	48235.50	4668.00	123089.90
	+528.72	42548.87	+3398.97	+4791.23	+561.36	10282.58
Samad	5471.20	26160.00	35172.00	29430.00	3552.00	99785.20
	+587.15	+2747.36	+3774.54	+3090.78	625.50	+8862.40
Anlay	3123.40	35352.00	20079.00	39771.00	2028.00	100353.40
	+304.88	+3108.34	+1959.94	+3496.89	+420.98	+7743.20
Korzok	6121.81	42840.00	39354.54	48195.00	1298. 18	137509.54
	+654.57	+3744.26	+4207.97	+4212.30	+235.430	+10754.97
Tsaga	2194.18	41192.72	14105.455	46341.81	5585.45	109419.63
	+661.99	+4385.23	+4255.65	+4933.39	+2193.93	+13367.00

Table 1.
Average income statement of each family from different livestock product (INR).

Figure 3.
Flock of Changthangi Pashmina Goat.

Figure 4.
Pashmina Goats at High Land Pasture.

for their livestock (**Figure 4**). The available pasture was sufficient for the number of livestock reared in Changthang by the Changpas till 1959, but during Chinese invasion in 1962 a lot of people from western Tibet migrated to Changthang side along with their livestock. They settled in Changthang and ultimately extra stocking of livestock on limited pastures caused damage to the pastureland due to overgrazing. Regarding pashmina production and its role in the economics of Changthangi people, the analyzed data shows that the highest average income at individual family level is fetched from Changthangi goats, which comes to around Rs 79,886.98, followed by Sheep Rs 33,412.88 and Yak, Rs 2621.63 annually. The overall average income of each family comes to Rs 116,021.49, in which Rs 79,886.98 is solely contributed by Changthangi goats, showing 68.85% of the total income annually. This result indicates that pashmina goats add to majority of the total income. The contribution of sheep is followed by pashmina goat represents 28.8% of the total income from livestock [1]. These results give a clear picture that the income generated from Changthangi goat in the form of Pashmina fibres, chevon, milk and manure forms the basis of their livelihood (**Figure 5**). In some villages like Koyul, Kharnak, Anlay, Korzok, Tsaga the income from pashmina goat contribute more than 90% of the total income.

2.1 Population status

The number of Changthangi goats in the year 2000 was 163,663 with a sudden increase to 223,093 in 2001. The populations remain static till 2004. In 2005 a sharp decline to 187,299 was noticed with marginal increasing trend to 208,611 in 2009 [6, 7]. The population of Changthangi goat was 217,771 in 2014 and has shown a slight increase to present 219,198 in 2019 [3, 8]. The Population status of pashmina is static for the past two decade, however the population of local goat (Malra) has drastically reduced from 68,838 in the year 2006 to 3162 in 2014 [8, 9] and now to slight increase to 16,199 in 2019 [3] which is a concern. The detail population status of pashmina goat is given in **Tables 2** and **3**.

Figure 5.
Milking of Pashmina Goats.

2.2 Diversity study

Genetic diversity study in Changthangi goats of Ladakh was done by this institute in 2020. The genetic distance analyzed based on the microsatellite allele variability between different clusters shows genetic nearness and farness between the Changthangi goats of different subpopulations, which could be very useful for

Category of animals	Yearwise number of heads/birds						
	2012–13	2013–14	2014–15	2013–16	2016–17	2017–18	2018–19
Sheep	105,241	103,477	10,8937	112,185	103,375	110,943	103,375
Non Pashmina Goat	3340	3162	2685	3126	19,099	161,99	190,99
Pashmina Goat	249,206	217,771	235,687	226,951	202,561	219,198	202,561
Total:-	357,787	324,410	347,309	342,262	325,035	346,340	323,035

Source: Sheep Husbandry Deptt.

Table 2.
Sheep and goat population as per livestock census/departmental survey.

S. No.	Item	Units	Physical target	Achievement up to ending December 2019
1	Wool	Kgs	101,670	102,009
2	Pasmina	Kgs	43,517	46,003.6
3	Mohair	Kgs	775	803
4	Mutton	Kgs	664,950	624,265.79
5	Milk	Litres	—	520

Source: Sheep Husbandry Deptt.

Table 3.
Sheep husbandry estimated production.

formulating future breeding plans in these goats. The Diversity study was done using highly polymorphic 15 microsatellites as recommended by FAO. The microsatellite are good candidates for breed characterization and diversity study. The study proved very useful for genetic investigations and assessing admixture in this goat populations. Bottleneck analysis revealed no recent bottleneck in Changthangi goats of Ladakh [10]. The strong inference that the Changthang breed of goat has not undergone bottleneck is important for goat breeders and conservationists, as it indicates that any unique alleles present in this breed may not have been lost. Therefore, it can be recommended that within breed diversity is actively maintained to enable these extensively unmanaged stocks to adapt to future demands and conditions and there is ample scope for further improvement in its productivity through appropriate breeding strategies. The substantial genetic variation and polymorphism observed across studied loci in the Changthangi Goats of Ladakh and the inference drawn from this study could be used for formulating future breeding plan and overall genetic improvement of this goat for Cashmere and meat production in whole of Ladakh.

2.3 Constraints and solutions

S. No	Constraints	Preferable solution
1.	Kid Mortality due to Hypothermia, undernourished, diarrhea of unknown etiology and Contagious Ecthyma.	Provision of Low cost Shelter for Kids Supplementary feeding of pregnant does and nursing does. Better health intervention by state sheep husbandry department.
2.	Difficulty in Data recording due to extreme hard terrain.	Provision of proper animal identification tags with data recorder under various govt schemes.
3.	Lack of scientific breeding plans.	Involvement of experts from research institutes and by adopting genetic diversity map developed by SKUAST-K Leh.
4.	Shortage of elite males (cruel method of traditional castration at 3 months of age) leading to inbreeding	Adoption of closed method of castration and culling of excess animal under proper guidance of culling committee.
5.	Health Problems like, CCPP, FMD, PPR Coccidiosis, Endo and ecto parasites.	Adoption of health calendar developed by our institute.
6.	Mass abortion of unknown etiology.	Brucella and other diseases needs to be rule out
7.	Lack of feed and fodder during long Winter season.	Provision of proper feed and fodder bank. Use of Complete feed, pelleted feed and silage feeding during winter developed by SKUAST-K Leh.
8.	Lack of proper shelter for Kids, pregnant and sick animals during winter season.	Provision of Low cost Shelter for Kids, pregnant and sick animals during winter season.
9.	Overgrazing and Pasture Degradation.	Strictly following age old tradition of rotational grazing. Preservation of winter grazing land.
10.	Wild attack by Tibetan wolf and Snow leopard.	Protection using latest infra-red light, iron mesh in collaboration with wildlife department.
11.	Hard living condition and lack of basic facilities to the Nomads.	Provision of basic amenities by Govt Department.
12.	Tourism a better avenue for the Local people.	Needs proper education of livestock rearer.
13.	Lack of Value Addition	Hands on Training programme on various livestock products processing.

S. No	Constraints	Preferable solution
14.	Marketing Problems.	Implementing Value Chain strategy from production to consumption with involvement of various stakeholders.

2.4 Future strategies

1. Breeding Plans like ONBS with Research Institute holding the nucleus herd or elite stock.
2. Selection of animals for optimum fiber production i.e lesser fibre diameter and longer staple length.
3. Proper Health Management (Dosing, Dipping and Vaccination)
4. Development of Disease Diagnostic kits.
5. Vaccine Production against CCPP.
6. Checking mass abortion and kid mortality.
7. Shelter management during peak winters.
8. Assisted reproductive technology for assisting the service of elite animals to more flocks.
9. Scientific optimum feeding during long winter (complete feed, pellet feed, feed block, silage feeding)
10. Establishment of Proper Feed Bank at each Cluster for winter storage of feed for livestock.
11. Pasture Development Programmes.
12. Value Addition of Pashmina fibres, Chevon, pelt and Goat Milk.

3. Changthangi sheep

Changthang region of Ladakh is the home tract for a potent dual purpose breed of sheep 'Changthangi' also known as "Changluk" locally. These breed of sheep is famous throughout the Ladakh region for its quality wool and mutton production (**Figure 6**). A total of about 113,554 of Changthangi sheep population sustain the economy of Changthangi people in the year 2003 which has drastically reduced to 49,654 in 2009 [7, 11]. Presently the population of Changluk is less than 70,000. This sudden reduction in the sheep population may be attributed to low income from sheep rearing compared to Pashmina goats, whose population is increasing year by year [12]. The low income return from sheep rearing is due to lay man approach for maintaining this breed as till date no specific breeding policy has been adopted for up gradation of

Figure 6.
Changthangi Sheep.

these sheep in the field and no effort has been made even by the state govt, which is clear from the fact that no such established farm of these sheep exist in the whole of Ladakh division. Low income from sheep is also attributed to religious belief as these people prefer other source of income from livestock other than meat production. A scientific approach in managing and rearing this Changthangi sheep is need of the time as the cost of rearing this animal is getting higher than its production. Also the required growth rate of the sector is possible only through selection based on important production traits. Among many of the constraints to sheep production in Ladakh, scarcity of feed, lack of breeding policy and high mortality has been the major limiting factors [13]. This is partly because sheep breeding in Ladakh is non-controlled, and health and nutrition management are very poor. Diminutive breeding efforts attempted as early as 1960s focused on crossbreeding of the indigenous breeds with exotic breeds (Russian Merino, Karakul) to improve growth and wool yield. Currently, exotic and crossbred's sheep in Ladakh constitute little proportion (<10%) of the total sheep population. The crossbreeding programme suffered from poor planning, not involving livestock owners and stakeholders in decision making and ownership of the initiatives on top of low regard to the potential of indigenous breeds. Studies made on Changthangi sheep breeds revealed within breed variation for growth and indicated feasibility for productivity improvement of indigenous sheep breeds through genetic means [12]. In the present situation owing to the importance of livestock in cold arid region of Ladakh, a framework for sheep breeding is seriously needed.

As per the recommendations of National commission on Agriculture from time to time, Northern Temperate Region of India including J&K has been earmarked for propagation of fine wool germ plasm. Big strides have been made in fine wool sector by the state. The production potential of the indigenous stock has been increased by more than double. However, till date Changthangi goat has been deprived of basic research for breed upgradation. Further, sheep breeding policy in the Union Territory of J&K and Ladakh should not be mono-cultured only for fine wool production because of its diverse agro- climatic conditions but it should be region specific and diversified.

3.1 Population status

On the contrary to the population trend of Pashmina goats in Ladakh the Changthangi sheep shows an initial sharp inclining trend from 66,822 in the year 2000 to 113,544 in 2004 with a subsequent declining trend to 49,652 in 2009 [7, 14, 15]. The population is somewhat static with 55,353 Changluk in 2014 [8]. However, the population of local Malluk has increased to 48,124 during the past few years making a total district sheep population of 103,477 in 2014. Presently the Changthangi Sheep population is 67,521 mostly restricted to Changthang region [3]. If suitable measures are not adopted, then the precious germplasm of Changthangi sheep may be lost during the coming years. The detail population status is given in **Table 4**.

3.2 Wool production

During our recent study at this institute, Changthangi sheep is one of the unique breeds having double coat with marked difference in fibre diameter and surface characteristics fibres. One of the striking observation reported in Changthangi sheep was that unlike other sheep breeds, Changthangi sheep produces double coat with marked difference in fibre diameter. The average fibre diameter of fine fibres was 14.35 ± 0.50 μm which was significantly lower ($p < 0.05$) than that of coarse fibres (40.04 ± 1.4 μm). The fibre diameter of fine fibres is comparable to that of pashmina growing from the Changthangi goats whose diameter range between 11–15 μm. The reason for presence of double coat and finer fibres may be attributed to the sub-zero temperature (even goes down up to −40°C) of the Changthang belt of Ladakh. Like Pashmina goats, Changthangi sheep might have adapted themselves to produce down

Category of animals	**Number of heads/birds**							
	2011–12	**2012–13**	**2013–14**	**2014–15**	**2015–16**	**2016–17**	**2017–18**	**2018–19**
Cattle	36,231	36,231	12,977	12,977	12,977	12,977	47,151	47,151
Buffaloes	—	—	—	—	—	—	0000	0
Dzo-Dzomoes	9495	34,147	34,174	34,174	34,174	34,174	18,877	18,877
Yark/Demos	13,420	18,877	18,877	18,877	18,877	18,877		
Horse/Ponies	—	5534	5534	5534	5534	5534	5534	5534
Mules	—	247	247	247	247	247	247	247
Donkies	—	5296	5296	5296	5296	5296	5296	5296
Camel	—	189	189	189	189	189	189	189
Dog		2949	2949	2949	2949	2949	2949	2949
Rabbits	—	80	80	80	80	80	80	80
Poultry	6093	20,829	20,829	20,829	20,829	20,829	20,829	20,829
Others (specify)	17,083	—	—	—	—	—	0	0

Source: Livestock Census Reports/departmental Survey 2012 After every 5 years.

Table 4.
Livestock population as per livestock census/Departmental survey.

fibres which help to protect them against the extreme cold conditions [16]. The production scale of finer fibre in Changthangi sheep should be studied further so that it can be used for improving the income of the livestock rearer of Ladakh. Till date the Changthangi sheep has been under rated for its quality wool production. The Quality of Changthangi sheep wool is categorized as carpet wool but our study indicates that it is medium type wool and have scope of good market by developing a fine textured apparels. Further, the undercoat produced from secondary follicle can be utilized for the development of fine textured fabrics with smooth and warmth which in turn will fetch more prices only next to pashmina.

3.3 Important features of breed

1. It is the only breed of sheep adapted to cold arid condition low hydric and hypoxic conditions with good returns to farmers (**Figures 6** and 7).

2. It has the longest staple wool fiber among the Indian sheep breed.

3. The breed is the tallest of Indian breed with very little or no health issues.

4. The breed produces the strongest wool among the Indian breeds.

5. The breed co-exists with pashmina goats for thousands of year and act as a protector of weak animals from severe winter when housed together.

6. Great variability is there in the body weights and thereby heralding great chances of improvement in meat production.

7. Variability does exist in wool production and wool quality, therefore, if required, these traits can also be set at desirable level (but within physiological limits of the breed).

Figure 7.
Changthangi Sheep on winter pasture.

8. Breed produces good quality mutton greatly relished by the local people of the area.

9. The breed produces secondary fibre with fibre diameter less than 13 μm.

10. The adult wt of the breed is 30–35 kg within 18–24 month making it a potential breed for mutton production.

3.4 Constraints and solutions

S. No	Constraints	Preferable solution
1.	Difficulty in Data recording due to extreme hard terrain.	Provision of proper animal identification tags with data recorder under various govt schemes.
2.	Lack of any breeding Policy	Formulation of breeding plan involving experts from concerned research institutes .
3.	Shortage of elite males (cruel method of traditional castration at 3 months of age) leading to inbreeding	Adoption of closed method of castration under proper guidance of culling committee.
4.	The low income returns from sheep rearing which is due to lay man approach for maintaining this breed.	Area specific Scientific sheep rearing practices to be developed. Educating local youth for profitable sheep farming.
5.	Reduce market value of wool.	Development of locally made woolen products.
6.	Reduce interest toward mutton production due to religious taboos.	Business strategy to be developed for sale of live animals.
7.	Health Problems like, Pneumonia, FMD, PPR Coccidiosis, Endo and Ecto parasites.	Adoption of health calendar developed by our institute.
8.	Mass abortion of unknown etiology.	Development of proper diseases diagnosis centre at each cluster.
9.	Lack of feed and fodder during long Winter season.	Provision of proper feed and fodder bank. Use of Complete feed, pelleted feed and silage feeding during winter developed by SKUAST-K Leh.
10.	Lack of proper shelter and feed for lambs, pregnant and sick animals during winter season.	Provision of Low cost Shelter for Kids, pregnant and sick animals during winter season. Supplementary feeding of pregnant does and nursing does.
11.	Overgrazing and Pasture Degradation.	Strictly following age old tradition of rotational grazing. Preservation of winter grazing land. Development of sound Pasture development programmes.
12.	Wild attack by Tibetan wolf and Snow leopard.	Protection using latest infra-red light, iron mesh in collaboration with wildlife department.
13.	Hard living condition and lack of basic facilities to the Nomads.	Provision of basic amenities by Govt Department.
14.	Tourism a better avenue for the Local people.	Coexistence of both tourism business and sheep farming. Needs proper education of livestock rearer.

DOI: http://dx.doi.org/10.5772/intechopen.103767

S. No	Constraints	Preferable solution
15.	Lack of Value Addition	Hands on Training programme on various livestock product processing.
16.	Marketing Problems for livestock products.	Implementing Value Chain strategy from production to consumption with involvement of various stakeholders.

3.5 Future strategies

1. Scientific breeding Policy.
2. Establishment of a nucleus breeding farm under a research institute to facilitate ONBS (3 tier system)
3. Establishment of Breed Society or Breed Cooperative like for Pashmina goats.
4. Proper Health Management (Dosing, Dipping and Vaccination)
5. Disease Diagnostic kits.
6. Checking mass abortion and kid mortality.
7. Shelter management during peak winters.
8. Scientific feeding during long winter (complete feed, pellet feed, feed block, silage feeding)
9. Establishment of Proper Feed Bank at each Cluster for winter storage of feed for livestock.
10. Pasture Development Programmes.
11. Value Addition of wool, Mutton and Pelt.
12. Pasture improvement programme
13. Shelter Management during winter
14. Better Marketing Route Map for wool.
15. Evaluation of Mutton quality
16. Diversity study using Molecular marker (Micro satellite)
17. Identification of Molecular Marker for production traits followed by Marker Assisted Selection (KAP, Myostatin and GH genes)
18. Introgression of Fec B Gene to increase twining in this sheep.

4. Yak

The yak species (*Bos grunniens*) represents a unique bovine species adapted to the Tibetan plateau of China and India at an altitudes of 3000 m above sea level where oxygen content is only 33% of that at sea level and intensity of ultraviolet radiation is 3–4 times that in lowland areas [17]. Consequently, yak adapted to this environment likely have special physiological mechanisms to protect their central nervous systems against hypoxic and oxidative injury. Semi-domestic Yak is commonly known as "the Ships of the Plateau" has been domesticated under hostile climates of high altitude in the Ladakh plateau and adjoining regions of greater Himalayas also (**Figures 8** and **9**). Apart from being pack and draught animal it provides milk, meat, fibre, hide. It is best

Figure 8.
Male Yak at High Land Grazing.

Figure 9.
Yaks at High Land Grazing.

known for its hardiness and resistant to extreme cold climate and hypoxic conditions. Yaks of Ladakh are very hardy and they rarely need any health care and its very rare to find any health issues in Yaks.

In India, Union Territory of Ladakh has the highest population of Yaks and its hybrids distributed in both Kargil and Leh districts. But, the National Research Centre on Yak was established at Dirang Arunachal Pradesh and till date Ladakh is devoid of any kind of support from NRC yak and Govt. of India leaving this precious germplasm to perish in times to come. The other areas, where Yaks are found include Drass valley and Doda district of Jammu & Kashmir, North-Western Himalayan region mainly Spiti Valley, Pangi Valley in Chamba and Sangla Valley in tribal district of Kinnaur, Pithoragarh district of Garhwal hills, Sikkim and Arunachal Pradesh. The base line data on the Yak of Ladakh is not available, whatsoever meager information available is based on few Yaks; therefore, status of real production, reproduction and other potential of this animal are wanting. The animal besides having agricultural utility has religious utility for the Buddhist community of Ladakh. No scientific improvement programme is in force to improve this animal, and the biggest hurdle in it is the lack of systematic and sufficient information about the species.

Yak when hybridize with domestic cow exhibit great degree of heterosis. The popularity of these hybrids among the farmer can be gauzed from the specific name to each back cross and hybrid. The real heterosis exhibited is not documented because of lack of information on potential of Yak.

In the cold arid zone of Ladakh the precious species of livestock viz. Yak (Bos *grunniens*) needs to be documented and improved by adopting the technologies which are already in vogue for cattle and buffalo improvement. To exploit the benefits of species hybridization by crossing Yak with cattle, the performance of different levels of inheritance of Yak with cattle germ-plasm needs to be studied in depth to harvest maximum benefits.

4.1 Population status

In India, the Yak population decreased drastically to about 30,000 in 1987 from 130,000 in the late seventies. The highest Yak population was about 21,400 in Arunachal Pradesh in 1972 and this population has shown sharp decline since & about 8921 Yaks had been estimated in 1997–98. These sharp declines in Yak population may be due to large scale cross breeding of Yak dams with local cattle to produce the hybrid Dzo and Dzomo, which is more useful and easily manageable than Yak due to its docile nature.

Likewise, in Ladakh also the same pattern of decline in Yak population has existed and at present it is very difficult to find elite Yaks in the population. The yak population has a sharp decline from 18,904 in 2007 to 13,420 in 2008 in Leh district similar pattern is also noticed in Kargil district [1, 18]. Presently the Yak population of Leh has slightly increased to 18,877 in 2019 [3]. However, the crosses like Dzo & Dzomo has increased abruptly from 9495 in 2008 to 34,174 in 2019 [3, 19]. This is mainly due to uncontrolled crossing with local cattle.

4.2 Constraints and solution

S. No	Constraints	Preferable solution
1.	Lack of Scientific Breeding and Management resulting in inbreeding.	Development of Breeding Policy is need of the time.

S. No	Constraints	Preferable solution
2.	Lack of any policy by govt organization for research and development.	Yak sub Centre needs to be established at Leh
3.	Lack of Knowledge about genetic diversity among different Yaks found in different areas of Ladakh.	Genetic characterization using molecular tools and genetic diversity study using microsatellite markers as suggested by FAO should be done. Breed registration should be done on priorty.
4.	Lack of Livestock Product development system	Trainings on Value addition. Promotion of Yak milk products like Yak cheese, yoghurt, butter etc
5.	Low income comparative to other Livestock	Development and marketing of Yak Meat, Pelt and wool which doesn't exist in Ladakh.
6.	Competition with Wild animals for grazing, shrinking of Pasture Land and Degradation.	Strictly following age old tradition of rotational grazing. Preservation of winter grazing land. Development of sound Pasture development programmes.
7.	Lack of proper nutritious feed during winter	Development and provision of Yak supplementary feed using locally available agriculture products.
8.	Indiscriminative breeding with local cattle	This should be checked and allowed until or unless warranted as per need.
9.	Most of the Elite stock goes to slaughter especially in winter.	Proper culling committee should be framed for culling the unproductive Yaks.
10.	Considered as Non Descriptive due to lack of data and basic study.	Genetic Characterization and registration as a Ladakhi Yak will be the first step in its conservation and improvement.

4.3 Future strategies

1. Genetic Characterization at molecular level of Yak found in different areas of Ladakh and its comparison with Yak found in those of China and rest of India.

2. Accordingly, adoption of Scientific Breeding Policy and Management System.

3. Value addition to a variety of Yak product, like milk, cheese, fibre, hair, pelt and bones.

4. Tourism oriented business like Yak safari.

5. Inclusion of Yak as a Pack animal for Army at border areas.

6. Based on the function of MT-III, it is predicted that MTs may play an important role for yak adaptation to the Tibetan plateau environment. Hence genes like Metallothionein-III (MT-III) Hypoxia HP1 gene in yak should be studied.

5. Zanskari horse

Zanskari and Chumurti breeds are the two main horse breed of Ladakh area, the Zanskari horse are found mainly in the Zanskar region of Ladakh and Chumurti being lesser and found mainly towards Chumur village of Changthang bordering China. The majority of the horses in Ladakh are Zanskari (**Figure 10**). These horses are very well adapted to run in hypoxic and freezing condition of Ladakh. The main utility of Zanskari horse is as a draught and sport animal. In some far flung village this horses are traditionally used for ploughing the agriculture field and for thrashing the crops (**Figure 11**). The famous traditional game of Ladakh 'Stapolok' (Polo) of Ladakh from ancient times is played using these horses. These horses are very hardy and was of great importance in earlier times for trades and transportation. In eighties owing to conserve this horse State Govt has opened a Zanskari horse breeding farm at Chuchot

Figure 10.
Zanskari Horse at Zanskar.

Figure 11.
Zanskari Horse at field work.

which is presently at critical stage due to lack of efficient breeding policy. Further, due to mechanization and development of road system, Zanskari horse, one of the precious germplasm adapted to the hypoxic conditions of high altitudes of Ladakh, is already endangered and needs immediate attention for its conservation. Though presently it has found its new way of income through tourism in trekking routes. Presently the population of Zanskari horse is 5534 which is almost equal to local donkey population of 5296 [3].

5.1 Constraints and solution

S. No	Constraints	Preferable solution
1.	Lack of Scientific Breeding and Management resulting in inbreeding.	Formulation of a sound Breeding Policy involving Research Institutes.
2.	Lack of any policy by govt organization for research and development.	Restructuring of present Zanskari Horse farms of Ladakh.
3.	Lack of Knowledge about genetic diversity among different Horses found in different areas of Ladakh.	Genetic characterization using molecular tools and genetic diversity study.
4.	Lack of use due to modernization	Introduction of this hoses in army, police and other departments.
5.	Low income comparative to other Livestock	Promotion of Polo sports. Tourism related sports trekking can increase their income.
6.	Lack of the Elite stock for propagation	Upgradation of Equine breeding centers in every blocks of Ladakh.

5.2 Future strategies

1. Breed Morphological, Physiological and biochemical Characterization of the Pure Zanskari horse breed which will be the first step towards formulating breeding policies and prioritizing the breeds for conservation in an effective and meaningful way.

2. Genetic Up-gradation of Zanskari Horse through Open Nucleus Breeding Scheme (ONBS).

3. Maintaining the population status of this breed of Horse to a safe level.

4. Physiological Studies to ascertain its performance and importance in cold arid region compared to other breed.

5. Provision of better Health and improvement in managemental practices.

6. Introduction of Artificial Insemination for genetic improvement in this breed.

7. To evaluate its present field of importance taking into consideration the present developed and changed scenario of Ladakh (promoting polo sport).

8. Strengthening of existing Zanskari Equine Breeding Farm AHD Leh,

6. Bactrian (Double Humped) camel

Another important livestock species of Ladakh region is Bactrian camel, which has played a great role in silk route trading transportation connecting China and Central Asia via Ladakh during 18th century. The present camel has been raised from a base population 18–22 camel since from 1937 when Ladakh was open for silk route trading,

Figure 12.
Double Humped Camel at Leh.

Figure 13.
Double Humped Camel Safari at Hundar.

since then no importation of Camel has been done due to border restriction. These camel had a critical population of hardly about 125 in 2008 [18] and presently increased to 189 [3]. In early 20th century these animals were abandoned by the local farmers due to lack of its use in day to day life. However, after opening of the Nubra village (Habitat) for tourism sector the business of Camel riding flourishes thereby increasing the economy of farmers. Presently, and Nubra valley becomes a centre of attraction for both national and international tourists (**Figures 12** and **13**). A total active business of camel riding for around 4 months from June to October was able to purchase feed and fodder for this camel for 12 months along with lots of extra savings to the Camel breeders. Though this Camel can be reared for milk, meat and wool, but the farmers rear it for tourism related business only, the reason being easy management and profitable income. It must be conserved on scientific lines with complete registration of all the available animals and maintenance of their breeding plan for extra income through its milk, meat and wool production.

6.1 Constraints and solution

S. No	Constraints	Preferable solution
1.	Still not recognized as any breed.	Genetic Characterization and Registration as a breed
2.	Lack of any policy by govt organization for research and development.	UT Ladakh administration should come up with a sound policy on conservation and improvement of Ladakhi Bactrian Camel. They can be used as pack animal for army personal at border areas.
3.	Lack of elite males	Importation of Elite male preferably from China or Mongolia is the only solution for improvement of this Camel.
4.	Lack of scientific breeding method leading to inbreeding	Trainings on Value addition. Promotion of Yak milk products like Yak cheese, yoghurt, butter etc
5.	Lack of proper housing even for mother & young ones and they live in open Snow (2–3 feet) covered plains during winter	Provision of proper housing especially during winter season.
6.	Lack of health facilities	A Camel health Centre should be established immediately in Hunder Nubra.
7.	Mortality of young ones	Provision of supplementary feeding, disease control and treatment can check the mortality of young ones.
8.	Lack of knowledge about Value addition of Camel Products	Development and marketing of Yak Meat, Pelt and wool which doesn't exist in Ladakh.
9.	Lack of Marketing for Camel product	Lack of Knowledge on camel product and processing
10.	Scarcity of feed and fodder (grazing land).	A state grazing land should be earmarked for Camel with provision of fodder production including seabuckthorn.

6.2 Future strategies

1. Genetic Characterization at Molecular level

2. Scientific Breeding Method.

3. Importation of male from Central Asia to check Inbreeding

4. Low cost Shelter especially for Nursing mothers and young ones and for winter seasons.

5. Introduction of feed processing equipment.

6. Identification and use of locally available feed resources.

7. Provision of Health facilities.

8. Methods to increase its use for agriculture.

9. Value addition of its products like milk, wool, pelt, bones etc.

7. Cattle

The local cattle which is a non-descriptive breed was registered as a Ladakhi Cattle breed recently in 2019 by National Bureau of Animal Genetics Resource Karnal, Govt of India. The Ladakhi Cattle is very well adapted to the hypoxic condition of Ladakh, it can thrive very well on meagre feed and highly resistant to most of the contagious disease (**Figures 14** and **15**). Certain reports on the local cattle indicate it to the Bos *taurus* species. In Ladakh the performance of Jersey crossed with these local cattle is very much appreciated by the farmers and more than 80% of the Ladakhi cattle are cross between Jersey and Local with an overall average milk production of 5–6 litres per day. There is complete trend of increase in cattle population of Leh district since from the year 1992 (24,836 Nos) to 2008 (36,231 Nos) leading to a sharp increase in total milk production of the district. The cattle population of Leh district in 2014 was 12977 with a static milk production. Presently there are 47,151 of Ladakhi cattle [3, 8].

Figure 14.
A Ladakhi Cow.

Figure 15.
Ladakhi Cattle.

This data reveals that after breed registration its number has increased drastically. The Ladakhi Cattle produce around 1–2 litres of milk (7–8% fat) per day with no input expenditure on feed and medicines and under improved managemental system its production goes to 5–6 litres of milk per day. The milk and colostrum of these cattle are explored for biomolecules having high medicinal values in humans. Presently, these local cattle which are adapted to Ladakh condition are being replaced by Jersey cattle which have many managemental and health issues regarding their adaptation in Ladakh. If this trend continues then this local genetic resource in future will be lost forever.

7.1 Constraints and solution

S. No	Constraints	Preferable solution
1.	Lack of any policy by govt organization for research and development.	UT Ladakh administration should come up with a sound policy on genetic improvement of Ladakhi Cattle involving breeders from research institutes.
2.	Low Production compared other breeds of the country.	Breed Improvement through establishment of a nucleus herd at Leh under a research institute.
3.	Indiscriminate breeding with other breeds	A complete check on unwarranted crossing with Jersey, Holstein and other breeds.
4.	Lack of adapted optimum producer Lack of elite males	Presently the exact picture production in this cattle is not known. A complete picture of different areas with production data may help in improvement programmes.
5.	Lack of scientific breeding method leading to inbreeding	Trainings on Value addition. Promotion of local cow milk products Yak cheese, yoghurt, butter etc
6.	Lack of proper housing system especially during during winter	Provision of proper housing during winter season.

S. No	Constraints	Preferable solution
7.	Lack of knowledge about Value addition of Camel Products	Development and marketing of Yak Meat, Pelt and wool which doesn't exist in Ladakh.
8.	Jersey as a better option for increased milk production.	Ladakhi Cattle can perform in higher altitude where Jersey breed cannot perform. Hence its propagation will be accepted in area where altitude is above 12000ft asl.
9.	Scarcity of feed and fodder especially during winter season.	Complete feed, pellet feed, silage feeding should be practiced using locally available agriculture produce like Barley, wheat, mustard, peas, their straws, maize and alfa alfa.
10.	Lack of research on milk and milk products	The milk and colostrum of high altitude livestock is considered to have special biomolecules, which may have a tremendous health benefits to the consumer.

7.2 Future strategies

1. A sound breeding policy (ONBS) to improve the production system of local cattle.

2. A planned study on the performance of various inheritance level of Jersey graded local cattle at Leh.

3. Study on resistance nature of native cattle against various infectious disease like FMD, Brucellosis, Mastitis etc.

4. Study on rumen flora so as to ascertain the reason for thriving on meagre feed resources.

5. Study and research on the milk parameters in detail.

6. Screening and Identification of novel biomolecules in milk and colostrum.

8. Poultry

Like other states of India with the pace of time there is great demand for poultry meat and eggs in Ladakh also, and all the demands are made by suppliers from Punjab and Kashmir. In the past there used to be a local poultry layer bird, very well adapted in every village house of farmers with broodiness. But due to lack of any policy the local birds of Ladakh has extinct and for the past 4–5 years it is very hard to find any local bird with broodiness. Though there is an increase in birds numbers mostly Vanraja Breed (dual purpose breed) from 6093 in 2007–08 and 20,829 in 2014 but the real poultry germplasm of the area is already lost. Normal Hatchery unit doesn't work well in Ladakh and hatching is a major problem in Ladakh condition due to cold, arid and high altitude of the region. Hence a hatchery suitable for Ladakh needs to be developed. Till date no research has been done to establish a broiler or layer line fit for cold arid zone of Ladakh, though DRDO claims to have done it but there is nothing to see at farmers/village level. Though a broiler business is not a profitable business in Ladakh condition but revival of age old backyard poultry has a great scope.

8.1 Constraints and solution

S. No	Constraints	Preferable solution
A	Broiler	
1.	Costly Housing	Designing of a special poultry house to maintain the required temperature with low cost and maintenance.
2.	Costly poultry feed (mainly due to long distance and costly transportation)	Production of Poultry feed using locally available agriculture produce.
3.	Poor growth rate	Improvement in room temperature and right feed.
4.	Difficulty in procurement of Day old chick	UT Administration should help in procuring day old chicks directly from Hyderabad or Chandigarh by air route.
B	Layers	
	Lack of suitable Parent stock	Government Institute should develop a Hatchery Unit in Ladakh itself.
	Lack of Hatchery Unit	Designing of a Hatchery Unit suitable for Ladakh Condition.
3.	Lack of Broodiness	A gene lost once is lost forever.
4.	Difficulty in procurement of Day old chick	UT Administration should help in procuring day old chicks directly from Hyderabad or Chandigarh by air route.
5.	Lack of proper housing design	A low cost poultry shed, semi underground type with mud bricks, one developed by KVK-Leh may reduce the housing cost and maintain required for birds

8.2 Other non-descriptive native livestock

1. *Purig sheep*: This is a small size sheep as described by Dr William Moorcroft the traveller who visited Ladakh in 17th Century (**Figure 16**). In his book he has described the origin of Ladakhi Sheep from the ancestor *Hunya* and Purig. The name 'Purig' has been derived from one of the tribe of Ladakh Purig whose ancestors are supposed to be originated from Western Tibet. The Purig sheep is famous among local people for its delicious mutton, though the mutton production is very less but it is greatly relished by local people during winter season and local festival. They also produce a small amount of wool which is used for making local woollen dress like sweater, socks, gloves, caps etc. This sheep are presently very few in numbers found in certain pockets of Sanjak, Sankoo and Panikhar blocks of Kargil district and Turtuk, Tyakshi and Thang Village of Leh District. Their population is estimated to be around 200–300 in numbers and is at the verge of extinction.

2. *Malluk sheep*: The Malluk name has been derived from local word meaning 'Local Sheep' is a medium, carpet wool type sheep found mostly in Lingshet, Photoksar, Skiu, Markha, Lamayuru and Khaltse area of Leh District and some parts of Zanskar block of Kargil district. The present sheep population is estimated to be

Figure 16.
Purig Sheep at Turtuk.

around 600–700 in numbers and declining every year due to crossing with Changthangi, Karakul and Marino Sheep. If proper action is not taken this germplasm may be lost in years to come.

3. *Malra goat*: The Malra name has been derived from local word meaning 'Local Goat', it is a medium size goat (**Figure 17**) commonly found in Khaltse

Figure 17.
Malra Goat at Lamayuru.

Lamayuru, Lingshet, Photoksar, Skiu, Markha, villages of Leh District. The goat produces a small amount of Cashmere fibre (50–100 gm/animal) and mainly used for chevon production. Their number is also declining and presently estimated to be around 700–800, and the figure of Malra given under the Census includes all types of non-pashmina goat breeds found in Leh.

4. *Chumurti like horse*: These breed resembles one kind of horse found in Lahul and spiti district of Himachal Pradesh hence the name Chumurti. These horsis are mostly found in Korzok, Koyul, Hanley and Chumur Villages of Ladakh. Their number is around 100–150 and needs proper census and study. This are medium size horses and mostly used by the nomads of Changthang for transportation of goods especially during migration from one pastureland to another.

5. *Ladakhi donkey*: In the earlier times, the Ladakhi Donkey was one of the most useful livestock for the Ladakhi people. Ladakh due to its mountainous terrain and lack of any mechanization was urgently in a need of transportation mechanism and this donkeys and yaks serve the purpose. The Ladakhi donkey was used for transportation of all the day to day needs, agriculture produces, fodders, equipment, water etc. Now presently with development of roads connecting every nuke and corner of Ladakh and with emergence of transport vehicle, its use has been restricted to some remote areas only, due to which people has started abandoning this donkeys. The lack of use of this species for domestic purpose has made it a stray animal resulting in reduction in its number. The present population is estimated to be around 3000 to 4000 in Leh district.

6. *Changthangi dog*: The Changthangi dog locally known as Changkhi is medium size hunting dog (**Figure 18**), thought to be a descendent of Tibetan Mastiff breed of dog from Tibet. They are found mostly in Changthang areas of Ladakh bordering China. This dogs are used by the nomads as a watch guard dog for their livestock on open pasture lands. They guard the livestock from wild animals attack like Tibetan wolf and snow leopard. They are very loyal to their masters

Figure 18.
Changthangi Dog.

The above mentioned livestock needs to characterised and conserved on priority and the first step in this is the registration of all these non-descriptive animals as proper Breed. Krishi Vigyan Kendra -Leh SKUAST-K in collaboration with National Bureau of Animal Genetics Resource-Karnal Govt of India has started the process of Characterization and Registration of all this precious germplasm in the year 2021 and it will be completed by the end of the year 2024.

8.3 Conclusions and recommendations

- Ladakh has been deprived of basic research facilities in animal science sector by the Central Govt for so many years due to which some of the best germplasm like Pashmina goats, Changthangi Sheep, Yak and Zanskari horse are declining in terms of numbers as well as in production.

- Beside of above mentioned animals the presence of Chumurti breed of horse, the Hunya sheep of Ladakh goes unnoticed and their status at present date is unknown. Hence, a strategy to bring these important breed back to normal tract is need of the time.

- Like population status, the production data from all these animals has remained static for the past so many years, the reason being lack of scientific intervention.

- Both the AHD and SHD are lacking basic facilities like animal disease diagnostic laboratory and treatment facilities and they should be strengthened.

- The present trend of Livestock population is not healthy and if necessary steps are not taken immediately a time will reach that all these precious animals will be lost forever.

- Though developmental work has been done by State Department of Animal Husbandry and Sheep Husbandry, there is an urgent need of State Research Institute and ICAR to come forward for undertaking basic research and conservation programmes on these animals.

- Intervention and Collaboration with CSWRI, CIRG, NRC Yak, NRC Camel and NRC Equine in coming days is also need of the time to preserve these precious genetic resources of Ladakh.

- One of the biggest hurdle in implementing livestock improvement programme in Ladakh is lack of true livestock population census. It is has been observed from present annual census data of Leh and Kargil district, that the concerned Department has never worked sincerely in generating true livestock population census. Due to which there is erratic population data and sometimes the total no of particular species remain constant for years together.

Acknowledgements

The author is very thankful to Indian Council of Agriculture Research for funding various research projects run by SKUAST-K in Union Territory of Ladakh. The author also acknowledge the contribution of SERB-DST Govt of India for funding the project genetic improvement of Changthangi Sheep.

Author details

Feroz Din Sheikh
Krishi Vigyan Kendra-Leh I, Sher-e Kashmir University of Agricultural Sciences and Technology of Kashmir, Leh, India

*Address all correspondence to: aizar22@gmail.com

References

[1] Sheikh FD, Azmat S, Ganai TAS, Misra SS, Azmat S. Socio-economic role of Pashmina goat on Changthang people. Indian Journal of Animal Science. 2008; **78**(11):1298-1301

[2] Ganai T.A.S., Kirmani M.A., Ganai N. A. and Tundup T. 2004; Pashmina production in Changthangi goats beyond the period of longest and shortest day. Proceedings of conference of Indian Society of Animal Genetics and Breeding-India

[3] Ladakh Autonomous Hill Development Council. 2019; District Census, Leh Ladakh.

[4] Von Bergen W. Wool Handbook. 3rd ed. Vol. Volume I. London, New York: John Wiley & Sons; 1963. pp. 343-365

[5] Sheikh FD, Bhattacharya TK, Misra SS, Kumar P, Sharma A. DRB3.2 gene polymorphism and its association with Pashmina production in Changthangi goat. International Journal of Immunogenetics. 2006;**33**:271-276

[6] Ladakh Autonomous Hill Development Council. 2005; District Census, Leh Ladakh

[7] Ladakh Autonomous Hill Development Council. 2009; District Census, Leh Ladakh

[8] Ladakh Autonomous Hill Development Council. 2014; District Census, Leh Ladakh

[9] Ladakh Autonomous Hill Development Council. 2006; District Census, Leh Ladakh

[10] Sheikh FD, Ganai TAS, Ganai AM, Alam S, Asmat S. Exploring genetic diversity in Cashmere goats of Ladakh for enhancing production. Small Rumina nt Research. 2021;**201**:106409

[11] Ladakh Autonomous Hill Development Council. 2003; District Census, Leh Ladakh

[12] Ganai TAS, Misra SS, Sheikh FD. Description of Changthangi Sheep of Ladakh. Indian Journal of Small Ruminants. 2010;**17**(1):32-40

[13] F.D. Sheikh, T.A.S. Ganai, T. Tsewang, S.S. Misra and S. Azmat, 2010; Changthangi sheep: A Valuable Genetic Resource. Agro-Animal Resources of Higher Himalayas, 365

[14] Ladakh Autonomous Hill Development Council. 2000; District Census, Leh Ladakh

[15] Ladakh Autonomous Hill Development Council. 2004; District Census, Leh Ladakh

[16] Malik AA, Khan HM, Sofi AH, Mir MS, Farooq J, Sheikh FD, et al. Wool characteristics of Changthangi Sheep. Small Ruminant Research. 2021;**199**: 106364

[17] Sheikh FD. Indian Wild Life Year Book. Dong: The Wild Yak of Ladakh; 2009. pp. 15-23

[18] Ladakh Autonomous Hill Development Council. 2007; District Census, Leh Ladakh

[19] Ladakh Autonomous Hill Development Council. 2008; District Census, Leh Ladakh

Chapter 4

Genetic Diversity and Evolution of Yunnan Chicken Breeds of China

Mohammed Alsoufi and Ge Changrong

Abstract

Chickens are the first type of bird that was domesticated and spread widely in the world to cover the growing demand for animal protein from meat and eggs, and it was cultivated from a wild ancestor known Red junglefowl (Gallus gallus). Yunnan Province is considered the most diverse in culture and biology among all the provinces of China. There are a total of more than 24 chicken breeds in Yunnan Province. These chickens are characterized by good quality of their meat and eggs, a good immune system against diseases, and the ability to adapt to various environmental and administrative conditions. Yunnan Province is one of the centers of domestication and evolutionary of chickens in the world. There are many studies that have been conducted to evaluate and study the genetic diversity and evolutionary relationship within and among chicken breeds in Yunnan Province and their relationship with wild chicken species and other chicken breeds using phenotypic markers, protein polymorphisms, SNPs marker, microsatellite marker, and mitochondrial DNA marker. However, there is no review that summarizes these studies, and most of these studies were authored in the Chinese language. Therefore, we have reviewed all studies that have been conducted on Yunnan chicken breeds diversity in Yunnan Province.

Keywords: genetic diversity, Yunnan chicken breeds, molecular studies, phenotypic studies

1. Introduction

Poultry at the present time is called the domesticated birds and is used in the production of meat and eggs, and includes the following—chicken, ostrich, duck, turkey, goose, quail, pheasant, peacocks, and guinea fowl [1, 2]. Chickens are the first type of bird that was domesticated and spread widely in the world to cover the growing demand for animal protein from meat and eggs and it was cultivated from a wild ancestor known Red junglefowl (Gallus gallus) regarding since more than eight thousand years [3, 4]. Currently, the scientific and technological advancement across the globe was reflected in the production and domestication of poultry, which led to the existence of many breeds resulting from breeding, genetic improvement programs, and natural selection. Moreover, the production system, modern chicken, and the focus of countries on obtaining the largest possible amount of animal protein with the lowest possible loss led to loss of genetic diversity, a decrease of genetic variation, and disappearance of many local breeds, so that many research centers have sounded the alarm about the necessity of preserving local chicken breeds as an indispensable genetic resource [5].

During the last decades, commercial chicken breeds were imported in into China, and then a crossbreeding process was carried out with local breeds to cover the growing demand for chicken products. The total number of chickens in China about 10 billion birds in 2015 representing all types of chicken, of which 44%, 37%, 9.5% and 9.5% broilers, yellow chicken, layer and hybrid broiler [6]. Although approximately 107 local chicken breeds have been registered in China by the research centers, most of them are raised in small groups and the rural area, and some of them are at risk, as there are about six breeds at risk of extinction and due to the hybridization process with commercial strains, the number of local breeds adapted to hard environmental conditions in Yunnan, like to the rest of China, is constantly declining, in addition to the decrease in genetic diversity, which will negatively affect the ability of local chickens to withstand harsh environments, resist diseases, and losing the characteristics of high-quality meat. From all the above, we note that it is needed to maintain the highest genetic variation of local breeds as a national genetic resource and globally, for the purpose of breeding and genetic improvement programs that we will need in the future [5, 7, 8]. In this article, we have reviewed all the studies that were conducted previously on the genetic variation of chicken breeds in Yunnan Province.

2. Yunnan Province as a center of animal domestication

Yunnan Province is located in Southern China, sharing borders with Myanmar, Laos, and Vietnam at (21°8′32″–29°15′8”N, 97°31′39″–106°11′47″E). Most of the Yunnan landscape is classed as a mountainous region with the Tropic of Cancer, which runs across the southern region. The province of Yunnan is an incredibly different geographical location that comprises mountains, valleys, lakes, and rivers. The climate in Yunnan ranges from the tropical oceanic monsoon in summers and dry interior monsoon in winters, combined with adequate sunshine, long frost-free periods, and abundant rainfall. In combination with these highly diversified geographic conditions, Yunnan exemplifies a vital biological diversity center of worldwide importance. Yunnan is the center of about half of China's varieties of greater vertebrate and plant types and various species of rare, widespread, and wild animals. Yunnan is also the territory for feral descendants of many species of livestock and has been suggested to be a center domesticated of prevalent animals, such as the pig, chicken, and the dog [3, 9]. Yunnan Province is considered the most diverse in culture and biology among all the provinces of China. Its varied environment, from snow-covered mountains to tropical environments, enabled it to possess many species of plants and animals that have no equivalent in the whole world. The wide range of topographical along with a tropical humidity has made Yunnan Province extremely diverse biology and with a high degree of endemicity of species, as it has become one of the richest areas in the world in terms of plant and animal resources with 17 thousand species of plants and the equivalent of the northern hemisphere combined [10]. Although the area of Yunnan Province does not exceed 4% of the total area of China, it contains about 42.6% of plants species protected and 72.5% of wild animals protected that are found throughout the country [11], and is also considered as a home of many animals, the most important of which is the South Asian Gorge, Indochina tiger, Asian elephant [12], box turtle, the Yunnan monkey [13], and red forest chickens species, in addition to it contains 11 national and nature reserves [14], moreover, Yunnan province has about 650 species of freshwater fish with 580 species are natives, this equates to 40% of freshwater fish in China [15]. Yunnan Province possesses many local poultry breeds with a large variety of various traits [16].

3. Yunnan chicken breeds and their phenotypic characteristics and location of domestication

According to the report of Yunnan provincial animal and poultry genetic resources committee [17], there are a total of more than 24 breeds, and these breeds are located and distributed across all regions of Yunnan Province (**Figure 1**), named Nixi chicken, Wuding chicken, Xishuangbanna game chicken, Chahua chicken, Dali chicken, Xichou black bone chicken, Yanjin black bone chicken, Daweishan mini chicken, Yunlong short led chicken, Yangbi Hang chicken, Dulong chicken, Lanping chicken, Taliu black bone chicken, Dehong chicken, Labai high leg chicken, Lanping silky chicken, Mengzi game chicken, Poya chicken, Tengchong white chicken, Wuliangshan black boned chicken (Puer feathered feet chicken, Nanjiang green and black boned chicken), Weixin chicken, Wenshan chicken, and Piao chicken. These chickens are characterized by the good quality of their meat, a good immune system against diseases, and the ability to adapt to various environmental and administrative conditions. Because of the introduction of commercial strains, the number of these local breeds is decreasing, therefore, measures must be created to conserve these genetic supplies [19]. Furthermore, there are many breeds living in villages and mountains in Yunnan Province that have not been recorded or unknown until now, according to Kun et al. [20], in their study, that has reported three new breeds; Frizzle chicken, Naked-neck chicken, and YN chicken (YN) in Nujiang Prefecture in Yunnan Province (**Table 1**). Moreover, these domestic breeds when compared to broiler or layer chicken breeds mostly still not yet extensively bred and selected and possess a poorer performance, therefore some of them are not financially useful as broilers and layer breeds. However, it will continue to be a resource of genetic materials for the reason that they have been synthetically selected and bred throughout a lengthy history of reproduction and breeding using standards and methods that are completely different from those used with commercial chickens breeding [29].

There are many breeds of chickens that have been formed during thousands of years in Yunnan Province, and they can be divided according to their production purpose into the following: Entertainment type, meat type, dual type, and eggs type [17, 18].

Entertainment type: These chicken breeds are similar to wild chicken breeds (Red Junglefowl) that still live wild in the forests of Yunnan Province and are characterized by their low production of eggs and meat and their small size (**Figure 2**).

Xishuangbanna game chicken: Entertainment type, an ancient breed with history 2000 years. Large body size (2.5 kg male and 1.7 kg female), the annual production of eggs is 100–120 egg, white skin, variable plumage color (mostly grayish-green mixed with golden red) long neck, small walnut comb, and red earlobe [30]. This breed is distributed in Xishuangbanna prefecture (Southwest of Yunnan Province), (21°09’N–22°36’N and 90°56′E–101°50′E, ~2429 m above sea

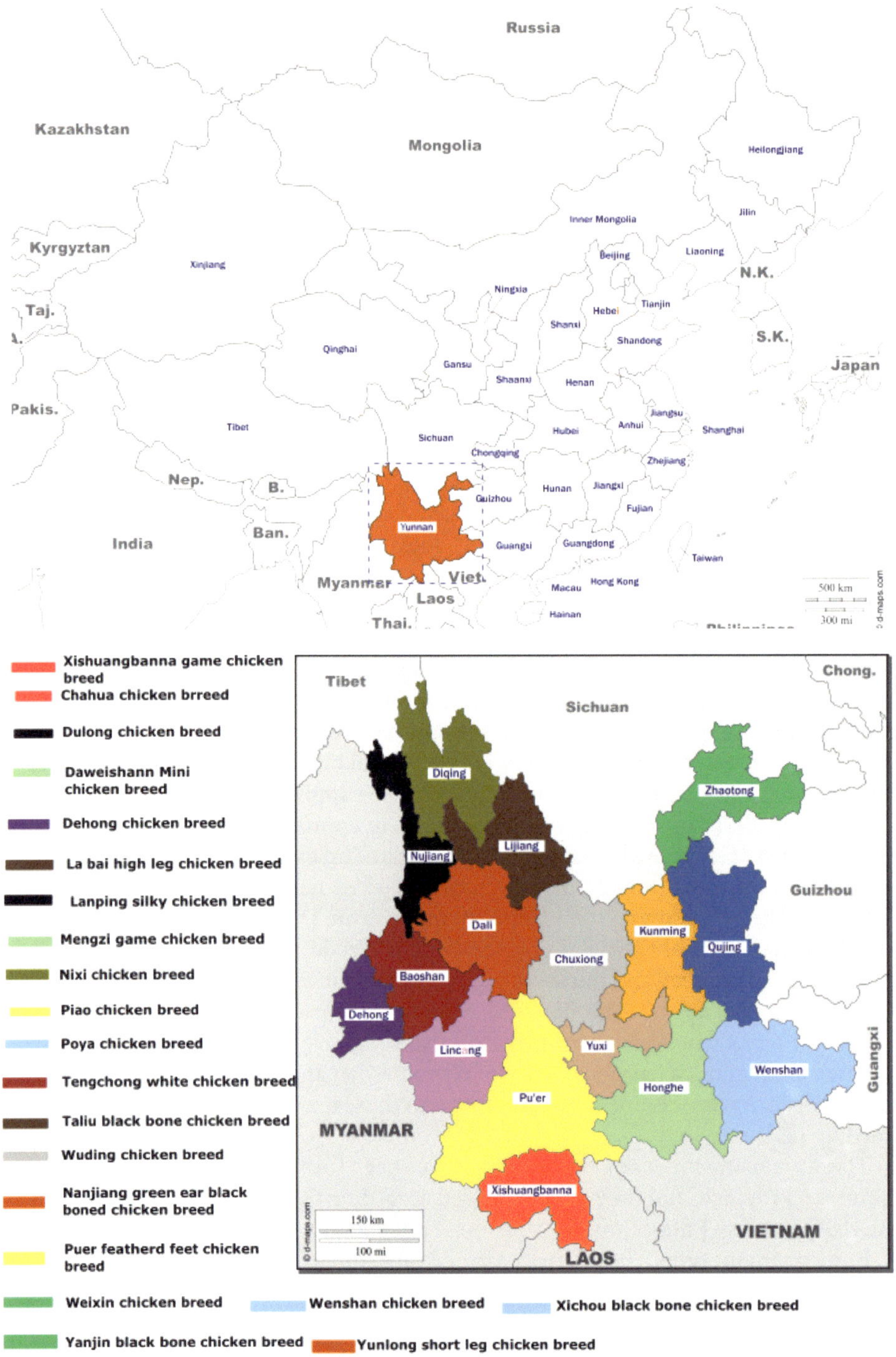

Figure 1.
Map of China and Yunnan Province indicating the location of domestication and distribution of chicken breed in Yunnan Province, [17, 18].

level); yearly average temperature 18.6–21.9°C, rainfall of 1200 mm–1700 mm, and humidity of 81–85% [17, 31].

Breed Name	Number of studied Samples	Haplotype Diversity	Nucleotide Diversity	References
Yanjin black-bone chicken	8	0.535	0.00103	Zhu et al. [21]
Tengchong chicken	45	0.881	0.01515	Liu et al. [3]
Wuding chicken	7	0.810	0.00919	Liu et al. [3]
Frizzle chicken	56	0.947	0.01268	Kun et al. [20]
Naked-neck chicken	56	0.938	0.01434	Kun et al. [20]
YN chicken	56	0.596	0.00239	Kun et al. [20]
Daweishan Mini Chicken	30	0.685	0.00443	Jia et al. [22]
Chahua chicken	30	0.476	0.00301	LU et al. [23]
Taliu black bone chicken	54	0.855	0.017	Miao et al. [24]
Lanping silky chicken	47	0.545	0.0046	Miao et al. [24]
Piao Chicken	50	0.883	0.015 03	Gongpan et al. [25]
Xichou chickens	36	0.884	0.01452	Huang et al. [26]
Yunlong short-leg chickens	80	0.851	0.01342	Ouyang et al. [27]
Dulong chicken	59	0.911	0.016	Li et al. [28]
Labai high-leg chicken	80	0.784	0.011 66	Ouyang et al. [27]

Table 1.
Haplotype diversity and nucleotide diversity in some Yunnan chicken breeds from previous studies.

Chahua chicken: Primitive type, good for running and flying, small body size (1.27 kg male and 1.07 kg female), yielding 100–140 eggs per year, the color of plumage is mixed gray, green, black, and yellow, the wattle and comb are red, and the skin color is white with some chicken is yellow. This breed is domesticated and distributed in Xishuangbanna and Licang, Puer, Dehong prefecture (Southwestern and Western Yunnan), (21°09'N–22°36'N and 90°56′E–101°50′E, ~2429 m above sea level); yearly average temperature 18.6–21.9°C ¬, rainfall of 1200 mm–1700 mm, and humidity of 81–85% [17, 23].

Dehong chicken: wild breed, small size (0.93 kg male and 0.67 kg female), the yielding eggs production is 8–12 eggs under natural conditions and 80 eggs per year in the farm. Single red comb, the skin color is white, the color of plumages is mainly mixed red with white, yellow with black, and black with white. This breed is mainly located and distributed in Mangshi, Longchun, and another county in Dehong prefecture (west of Yunnan Province), (23°50'N–25°20'N and 97°31′E–98°43′E, ~893 m–1200 m above sea level); rainfall of 1400 mm–1700 mm [17].

Mengzi game chicken: Entertainment type, large body size, tall, thick bones and strong (3 kg male and 2 kg female), the annual production of eggs is 50–80 egg, hard feeding, and strong resistance to diseases. White yellow skin color, red meat color, variable plumage color (mainly dark green, red with black, yellow with black) long neck, small walnut comb, and red earlobe. This breed is located and lived in Mengzi

Figure 2.
The morphology of entertainment type native breeds in Yunnan Province in China, [17, 18].

County, Honghe prefecture (Southeast of Yunnan Province), (23°01'N–23°34'N and 103°13'E–103°49'E, ~200 m–2567 m above sea level); yearly average temperature 18.6°C, rainfall of 1200 mm–1700 mm, and humidity of 72% [17].

Dulong chicken: Dual type with minimal production costs, hard feeding, and strong resistance to diseases. Small body size (0.97 kg male and 1.15 kg female), the yielding eggs production is 55–75 eggs. The skin color is white, the color of plumages is mainly mixed red with white, yellow with black, and black with white. This breed is distributed in Gongshan County, Nujiang prefecture (Northwest Yunnan), (27°40'N–28°45'N and 98°45'E–98°30'E, ~4964 m above sea level); yearly average temperature 16°C, rainfall of 2932 mm–4000 mm, and humidity of 90% [32].

Daweishan mini chicken: This breed is a slow growing, small size (0.85 kg male and 0.68 kg female), aggressive, pectoral muscles, and thighs are combined with strong, thin bones, yielding about 60 eggs per year, The color of plumages are mainly white, yellow, red flowers, the comb is red and multiple [27]. This breed is distributed in Pingbian County, Honghe prefecture (Southeastern of Yunnan Province), (22°49'N–23°23'N and 103°24'E–103°58'E, ~154 m–2590 m above sea level): yearly average temperature 16°C, rainfall of 1450 mm–1700 mm, and humidity of 33.5–80.9% [17, 22].

Figure 3.
The morphology of eggs production type native breeds in Yunnan Province in China, [17, 18].

Egg production type: These breeds are mostly characterized by their small size and producing many eggs in year (**Figure 3**), which are the following:

Nixi chicken: Egg production type, its egg production reaches 156–221 eggs annually, the body size is small (1.6 kg male and 1.2 kg female), single red comb, long tail, gray shank, most of them white skin and some are black, variable plumage (mixed red and yellow and black for male and most of the black color with some white color for female and mixed yellow with black color). The regions of domestication and distribution of this breed are lives and are Shangri La County, Diqing prefecture county (Northwestern Yunnan), (26°52'N–28°52'N and 99°23′E–100°19′E, ~2800 m above sea level); yearly average temperature 7.4–13.5°C, rainfall of 270 mm–500 mm [16, 17].

Yunlong short-leg chicken: Yunlong chicken is a type of eggs production (160–190 eggs per year), small size (1.9 kg male and 1.6 kg female), good meat quality, unique flavor features, strong adaptability, the skin color is black or white and the color of plumages are mainly red and yellow. This breed is located and lives in Yunlong County, Dali prefecture (Central Yunnan plateau), (25°28'N–26°23'N and 99°52′E–99°46′E, ~730 m–3663 m above sea level); yearly average temperature 15.9°C, rainfall of 730 mm [17, 27].

Poya chicken: This breed is a type of eggs production (150–210 eggs per year), small size (1.4 kg male and 1.3 kg female). This breed has a red and single type of comb, the plumage color is red, white, black, and yellow. The skin color is white. Funingxian County, Wenshan prefecture (Southeast of Yunnan province), is the location of domesticated of this breed, (23°41'N–23°52'N and 105°53′E–106°10′E, ~823 m above sea level); yearly average temperature 19.3°C, rainfall of 1200 mm [17].

Dual type: These breeds are of medium size, and their production of eggs and meat is acceptable (**Figure 4**), and the most important of them are the following:

Figure 4.
The morphology of dual production types native breeds in Yunnan Province in China, [17, 18].

Xichou black bone chicken: Dual-type, aggressive and vigilant, bodyweight is medium (2 kg and 1.7 kg for female), the number of eggs produced is about 100–130 annually, the shank, skin, and bone are black, a varied plumage color, comb, wattle, earlobe, and beak are red and black. This breed has domesticated and distributed in Xichou County, Wenshan prefecture (Southeastern of Yunnan Province), (23°05'N–23°37'N and 104°22′E–104°58′E, ~667–1962.9 m above sea level); yearly average temperature 15.5°C, rainfall of 1100 mm–1600 mm, and humidity of 78% [16, 17].

Wuding chicken: Dual type (meat and eggs production). The body size of this chicken breed has small and large types (3 kg for male large type, 2.1 for male small type, and 1.7 kg for female), the number of eggs produced reaches 90–130 eggs per year. The color of plumages, skin, and bone is varied, the comb is red and single. Some of these breed chickens have feathered feet. This breed is located in Wuding

County, Chuxiong prefecture (Central Yunnan plateau), (25°19'N–26°11'N and 101°56'E–102°29'E, ~862 m–2956 m above sea level); yearly average temperature 15.1°C, rainfall of 959 mm [16, 17].

Piao chicken: The common name is Piao chicken, this breed has no pygostyle, tail bones, tail fat gland, tail feathers, caudal vertebra, and uropygial gland. Medium size, the bodyweight is reached to 2 kg for male and 1.7 kg for female, the number of eggs produced reaches 100–130 eggs per year, the skin, meat, and shank color mostly black and some skin is white, a varied plumage color (reddish-brown, black, white, yellow with flower color, single red comb. The location and distributions of this breed are Zhenyuan County, Puer prefecture (Southwestern of Yunnan), (23°24'N–24°22'N and 100°21'E–101°31'E, ~774 m–3137 m above sea level); yearly average temperature 18.5°C, rainfall of 1284 mm, and humidity of 78% [17, 33].

Taliu black bone chicken: Dual type (meat and eggs production). Large size (2.4 kg male and 2 kg female), the yielding eggs production is 90–120 eggs, with strong black skin, black bones, black meat, good meat quality, plumages colors are two types, red mixed with yellow, black color and white color. The location of this breed is Yongsheng County, Lijiang prefecture (Northwestern Yunnan), (25°59'N–27°04'N and 100°22'E–101°11'E, ~2890 m above sea level); yearly average temperature 13.5°C, rainfall of 936 mm [17, 24].

Lanping silky chicken: Dual type (meat and eggs production), medium size (1.9 kg male and 1.6 kg female), the production of eggs is about 110–120 eggs annually, the meat flavor is superior and unique, the tail is very short, the colors are mainly red meat, white skin, black shank, the plumage color is yellow with black tail and wings. This breed is located and lives in Lanping County, Nujiang prefecture (Northwestern Yunnan), (26°06'N–27°04'N and 98°58'E–99°38'E, ~1350 m–4435 m above sea level); yearly average temperature 13.7°C, rainfall of 1002 mm [17, 24].

Tengchong white chicken: Dual type (meat and eggs production), the body size is a medium (2.1 kg male and 1.7 kg female). The production of eggs reaches 100–150 eggs per year. This breed has a red and single type of comb, the plumage color is white. The color of shanks, skin, bone is black. This breed is distributed in Tenchong County, Baoshan prefecture (western Yunnan), (24°38'N–25°52'N and 98°05'E–98°46'E, ~930 m–3780 m above sea level); yearly average temperature 14.8°C, rainfall of 1469 mm, and humidity of 81.7% [16, 17].

Nanjiang black-boned chicken: Dual type (meat and eggs production), the body size is a medium (2.9 kg male and 2.2 kg female). Small head, green ear, the color of skin, meat, and bone is black. This breed has a red and single type of comb, the plumage male color is red, in addition to whit color, the female color manly is yellow and some chicken color is white. This breed is located and lives in Wuliangshan County, Dali prefecture (Western Yunnan province), (24°32'N–25°10'N and 100°06'E–100°41'E, ~994 m–3061 m above sea level); yearly average temperature 19.2°C, rainfall of 770 mm, and humidity of 68% [17].

Meat production type: These breeds are mostly characterized by their large size and high efficiency in the production of meat (**Figure 5**), which are the following:

Yanjin black bone chicken: Meat production type, large body size (3.1 kg male and 2.4 kg female), yielding 120–160 eggs per year. The skin, eyes wattle, face, ear, and comb are black, also the beak, toes, and shanks are black, a variable plumage color (mostly white or black), the location and distribution of this breed are Yanjin County, Zhaotong prefecture county (Northeast of Yunnan Province), (26°34'N–28°40'N and

Figure 5.
The morphology of meat production types native breeds in Yunnan Province in China, [17, 18].

102°52′E–105°19′E, ~267 m–4040 m above sea level); yearly average temperature 6.2–21°C, rainfall of 1100 mm [16, 30].

Labai high leg chicken: Meat production type, the meat is tender and unique, tall body, long greenshank, large size (2.7 kg male and 2.3 kg female), the yielding eggs production is 90–120 eggs, the skin color is white, and the color of male plumages are mainly yellow, with black wings and tail and yellow and black for female. Ninglang County, Lijiang prefecture (Northwestern Yunnan), is the region of domestication and distribution of this breed, (26°36'N–27°56'N and 100°22′E–101°15′E, ~1350 m–4510 m above sea level); yearly average temperature 12.7°C, rainfall of 918 mm, and humidity of 69% [17, 27].

Puer feathered feet chicken: Meat production type, this breed has a feather on feet, large size (3 kg male and 2.4 kg female), the production of eggs reaches 90–130 eggs per year. The skin color is mostly black and some of them are white, the plumage color is reddish-brown or yellowish-brown, and the color of the tail is black, in addition to some females is white or black. This breed is located and distributed in Puer County, Puer prefecture (Southwestern of Yunnan), (23°56'N–24°29'N and 100°22′E–101°15′E, ~795 m–3371 m above sea level); yearly average temperature 18.3°C, rainfall of 1086 mm, and humidity of 77% [17].

Weixin chicken: Meat production type, the bodyweight is large (4.3 kg for male and 3.5 kg for female), The production of eggs reaches 128 eggs per year. The legs are tall and thick, this breed has a red and single type of comb, white skin, the plumage color mostly is a black female and red or red with black for male. Weixin County, Zhaotong prefecture (Northeast of Yunnan Province), is the location and distributions of this breed, (27°42′30"N–28°07′30"N and 104°41′15″E–105°18′45″E, ~480 m–1902 m above sea level); yearly average temperature 13.3°C, rainfall of 1060 mm, and humidity of 84–89% [17].

Wenshan chicken: Meat production type, large size (2.6 kg male and 2 kg female). The production of eggs reaches 90–120 eggs per year. This breed has a red and single type of comb, the plumage color is brown for female and reddish-brown for male. This breed is located and has domesticated in Wenshan city and Maguan County, Wenshan prefecture (Southeastern of Yunnan Province), (22°34'N–24°28'N and 103°30′E–106°11′E, ~123 m–2991 m above sea level); yearly average temperature 15.8–19.3°C, rainfall of 1224 mm, and humidity of 76.7-86% [17, 18].

4. Genetic diversity molecular studies of Yunnan chicken breeds

Yunnan Chicken breeds are mainly reared in remote mountainous areas. But what is disturbing is that these breeds are less affected by the outside world, but it is being bred in a traditional selection by farmers, nevertheless, in recent periods, the genetic resources of poultry have received strong support from the Ministry of Agriculture and the provincial government to preserve these genetic resources and expand production for many breeds [34].

During the past 30 years and assuming that Yunnan Province is one of the centers of domestication and evolutionary of chickens in the world, there are many studies that have been conducted to evaluate and study the genetic diversity and evolutionary relationship within and among chicken breeds in Yunnan province and their relationship with wild chicken species and other chicken breeds using phenotypic markers [35], protein polymorphisms [36], and mitochondrial DNA marker [20, 22–24, 27, 31, 37].

4.1 Mitochondrial DNA marker studies

Jia et al. [22] studied the origin and genetic variation of 30 Daweishan Mini chick ens breed and compared them with five species of red jungle fowl (G.g. bankiva, G.g. gallus, G.g. murghi, G.g. jabouill, and G.g. spadiceus) sequence that downloads from previously published data (GenBank) using mtDNA. They identified 18 variable sites and observed six haplotypes. Their conclusion has indicated multiple origins for the Daweishan breed and the subspecies G.g. spadiceus is more contributed of Daweishan Mini chickens breed evolution. Similarly, the conclusion has obtained by LU et al. [23] revealed that multiple origins for Chahua chicken and Daweishan breed, and the subspecies G.g. spadiceus is more contributed of Daweishan Mini chicken breed and Chahua chicken breed evolution. This study has conducted on 30 Daweishan Mini and 30 Chahua chicken breeds to assess the origin and genetic variety of these breeds using mtDNA marker.

Furthermore, Gongpan et al. [25] in their study that conducted on 50 chickens of Piao chicken breed to evaluate the genetic diversity. The results revealed that the genetic variation of this breed is high and has multiple origins for the Piao chicken breed. However, a study has been conducted to investigate the relationship of Dulong chicken breed with Chinese chicken breeds (Pengxian chickens, Jinyang chickens, Emei chickens, Jiuyuan chickens, Muchuan chickens, Miyi chickens, Shimian chickens, and Tianfu chickens) by using mtDNA analysis. The results have revealed the close relationship between Dulong chickens and studied chicken breeds and have suggested that Dulong chickens have a single matrilineal lineage [28].

Ouyang et al. [27] studied the genetic variation of 257 individuals from three chicken breeds (Labai high-leg chicken breed, Daweishan mini chicken breed, and Yunlong short-leg chicken breed) and using mtDNA) D-loop sequences. Based on genetic diversity results, they have suggested that there is a rich genetic diversity in these studied breeds and all of these breeds have a multiple maternal lineage, which supports the concept of various maternal ancestry of chicken. Wang et al. [38] used the mtDNA D-loop sequence to explore the origin and genetic variation of 30 individuals from the Labai high-leg chicken breed. Haplotype diversity and nucleotide diversity were 0.763 and 0.031, respectively. The findings showed that this breed kept a wealthy genetic variety and multiple origins for the Labai high-leg chicken breed.

4.2 Microsatellite's markers studies

There are many of studies have been done by using microsatellites markers and most of these studies indicated to increase of genetic diversity within population of Yunnan chicken breed. Huo et al. [16] examined the genetic diversity and association between seven native breeds in Yunnan Province (Tengchong chicken, Banna chicken, Nixi chicken, Chahua chicken, Wuding chicken, Yanjin chicken, and Xichou chicken,) and Red Junglefowl chicken by utilizing 28 microsatellite markers. The numbers of alleles that were identified 342, 121 of them were specific, the heterozygosity among the population was high (0.663) and the FIS was low (−0.098–0.005) indicating the weakness of inbreeding between population, the FST, and the distance of genetic were high (0.1757–0.3015) and (0.4232–0.6950) respectively, indicating the high diversity among populations. Li et al. [39] also have used 30 microsatellite markers to study the genetic diversity of six Yunnan chicken breeds (Chahua, Xishuangbanna game, Wuding, Yunlong short-legged, Yanjin silky, and Tengchong chicken). The results have demonstrated the lowest value of heterozygosity was in Wuding chicken, and the highest value was in the Tengchong breed. In addition to the Yunlong chicken, Yanjin and Wuding chicken were grouped together, while Tengchong chicken and Xishuangbanna chicken were grouped together, however, Chahua chicken had grouped alone.

Ye et al. [40] used 33 microsatellites to evaluate the population structure and genetic diversity of 30 individuals from the Chahua chicken breed. The results showed an increased average value of heterozygosity (0.6129) and the polymorphism information content (0.5276) for all experimented microsatellites. LangThui et al. [41] used 33 microsatellites to assess the level of genetic diversity of 50 individuals from the Nixi chicken breed. The results have indicated to increase in the genetic diversity of this breed with average heterozygosity of 0.63 and 0.551 for polymorphic information content. Jia et al. [42] used 33 microsatellites to evaluate the population structure and genetic variability of 30 individuals from the Daweishan Mini chicken breed. The results have indicated to decrease in the heterozygosity (0.1737) and the polymorphism information content (0.3279) with increased homozygosity. Qian et al. [43] investigated the genetic diversity and genetic structure of 53 unrelated individuals from the Wuding chicken breed by using 25 microsatellite markers. The results indicated that the genetic diversity of this breed is high with average heterozygosity of 0.6957 and 0.6382 for the polymorphism information content of 25 microsatellites that have been studied. Chen et al. [44] examined the genetic diversity of 53 individuals from the Yanjin black-bone chicken breed, the findings revealed that the Yanjin black-bone chicken breed was rich in genetic variation with average heterozygosity of 0.6232 and 0.5712 for polymorphism information content.

4.3 Microsatellite's markers studies

Wang et al. [45] have done an experiment on 10 chickens of Dulong Chicken breed to study the population structure and genes selection in the period of chicken domestication based on whole-genome resequencing using single nucleotide polymorphisms marker and the approach of fixation index. The results have identified 18,262,807 SNPs from the 10 genomes of Dulong Chickens and five genomes of Red Jungle Fowls that have been downloaded from the NCBI. The findings also have obtained 469 candidate genes, these genes may be related to small size, aggressiveness, and disease resistance in Dulong Chickens. Moreover, Guo et al. [46] used two methods (heterozygosity and fixation index) based on whole-genome resequencing to investigate selection signatures between eight individuals of Xishuangbanna chicken breed genome and six of Red Jungle Fowl genome that have been downloaded from the EMBL-EBI database. The results have identified more than 16 million SNPs from all individual's genomes and identified 413 candidate genes that are related to energy metabolism, disease resistance, aggressive behavior, immunity, and growth.

In addition, many studies have included some of Yunnan chicken breeds as a reference breed or experimented breed, particularly, chicken breeds that still preserve their morphological characteristics (Chahua chicken breed, Xishuangbanna game chicken breed, Dehong chicken breed, Mengzi chicken breed, Daweishan mini chicken breed, and Dulong chicken breed) which are similar to chicken species that wildly living in south Asia, including Yunnan Province to understand evolutionary of chicken, genetic variation and genetic relationship between them and other Chinese domesticated breeds, global domesticated chicken breeds, wild species chicken, commercial chicken breeds using SNPs marker [47–51], copy number variants [52], microsatellites marker [53], and mtDNA marker [54, 55].

5. Conclusions

The ecological and topographical diversity of Yunnan Province in China has been reflected in the biological diversity, as this region is considered one of the centers of genetic resources for living organisms. Including chicken, as it is one of the centers of domestication of chickens in the world where it contains all types of chicken breeds—fancy breeds, meat breeds, and eggs breeds. Research centers have begun to conduct many studies on the genetic variation and evolution of chickens using molecular methods. As many breeds are still discovered in succession to this day, it is necessary to make more efforts to enumerate and describe all the chicken breeds that exist in Yunnan Province.

Conflict of interest

The authors declare no conflict of interest.

Author details

Mohammed Alsoufi[1,2]* and Ge Changrong[1]

1 College of Animal Science and Technology, Yunnan Agricultural University, Kunming, Yunnan Province, People's Republic of China

2 Faculty of Agriculture, Department of Animal Production, Sana'a University, Sana'a, Yemen

*Address all correspondence to: m-alsofy@hotmail.com

References

[1] Aho PW. The world's commercial chicken meat and egg industries. In: Commercial Chicken Meat and Egg Production. Boston, MA: Springer; 2002. pp. 3-17

[2] Roenigk WP. Muscle growth and development. Keynote address: World poultry consumption. Poultry Science. 1999;**78**:722-728

[3] Liu Y-P, Wu G-S, Yao Y-G, Miao Y-W, Luikart G, et al. Multiple maternal origins of chickens: Out of the Asian jungles. Molecular Phylogenetics and Evolution. 2006;**38**:12-19

[4] Aho P. Challenges and opportunities for marketing poultry products in developed and developing countries. Presented on behalf of The American Soybean Association in Romania and Turkey; 2004

[5] Chang C-S, Chen C, Berthouly-Salazar C, Chazara O, Lee Y, et al. A global analysis of molecular markers and phenotypic traits in local chicken breeds in Taiwan. Animal Genetics. 2012;**43**:172-182

[6] Zheng MQ, Gong GF, Gao HJ. The monitoring report of poultry industry development in 2015. China Poultry Science. 2016;**38**:67-70

[7] Chen. Animal Genetics Resources in China. China Agricultural Press; 2012

[8] Vaarst M, Steenfeldt S, Horsted K. Sustainable development perspectives of poultry production. World's Poultry Science Journal. 2015;**71**:609-620

[9] Pang J-F, Kluetsch C, Zou X-J, A-b Z, Luo L-Y, et al. mtDNA data indicate a single origin for dogs south of Yangtze River, less than 16,300 years ago, from numerous wolves. Molecular Biology and Evolution. 2009;**26**:2849-2864

[10] Liang L. Biodiversity in China's Yunnan province. Biodiversity and Conservation. 2011;**23**:2012

[11] Yang Y, Tian K, Hao J, Pei S, Yang Y. Biodiversity and biodiversity conservation in Yunnan, China. Biodiversity and Conservation. 2004;**13**:813-826

[12] Smith AT, Xie Y, Hoffmann RS, Lunde D, MacKinnon J, et al. A Guide to the Mammals of China. Princeton University Press; 2010

[13] Blanck T, Zhou T, McCord W. The Yunnan box turtle, Cuora yunnanensis (Boulenger, 1906); historical background and an update on the morphology, distribution and vulnerabilities of the only known living specimens. Sacalia. 2006;**13**:14-35

[14] Grueter CC, Jiang X, Konrad R, Fan P, Guan Z, Geissmann T. Are Hylobates lar extirpated from China? International Journal of Primatology. 2009;**30**:553-567

[15] Chen X-Y. Checklist of fishes of Yunnan. Dongwuxue Yanjiu. 2013;**34**(4):281-343

[16] Huo J, Wu G, Chen T, Huo H, Yuan F, et al. Genetic diversity of local Yunnan chicken breeds and their relationships with Red Junglefowl. Genetics and Molecular Research. 2014;**13**:3371-3383

[17] YPAPGRC. Yunnan provincial animal and poultry genetic resources committee, Animal genetic resources in yunnan. Kunming: Yunnan science and technology press; 2015

[18] NAPGRC. National Animal and Poultry Genetic Resources Committee, China's livestock and poultry genetic resources. Beijing: China Agriculture Press; 2011

[19] Chen G-H, Wang K-H, Wang J-Y, Ding C, Yang N, et al. Poultry genetic resources in China. Shanghai, China: Shanghai Scientific and Technological Press; 2004

[20] Kun W, Tengfei D, Qihua L, Mingzheng L, Fengyun L, et al. Genetic Diversity analysis of three newly discovered native chicken breeds in Yunnan using mtDNA D-loop region sequences. China Animal Husbandry and Veterinary Medicine. 2020;**47**:3232-3241

[21] Zhu W, Li H, Wang J, Shu J, Zhu C, et al. Molecular genetic diversity and maternal origin of Chinese black-bone chicken breeds. Genetics and Molecular Research. 2014;**13**:3275-3282

[22] Jia X, Tang X, Lu J, Gu R, Ge Q, et al. The investigation of genetic diversity and evolution of Daweishan mini chicken based on complete mitochondrial DNA D-loop region sequence. Journal of Nanjing Agricultural University. 2015;**38**:656-660

[23] Lu J, Jia X, Tang X, Fan Y, Tang M, et al. Genetic diversity of two local Yunnan chicken breeds and their relationships with red jungle fowl. Journal of Zhejiang University (Agriculture and Life Sciences). 2016; **42**:385-390

[24] Miao Y, Sun L, Wang S, Ouyang Y, Zhang Y, et al. Genetic diversity analysis of three indigenous chicken populations in Yunnan using mtDNA D-loop sequences. Journal of Yunnan Agricultural University. 2015;**30**:395-401

[25] Gongpan P, Liu L, Li D, Yuan Y, Sun L, et al. The investigation of genetic diversity of Piao chicken based on mitochondrial DNA D-loop region sequence. Journal of Yunnan Agricultural University. 2011;**26**:211-223

[26] Huang D, Li D, Yuan F, Lin C, Li H, Su S. Genetic diversity of mitochondrial DNA D-loop region of Xichou black-bone chicken. Journal of Yunnan Agricultural University. 2010;**25**:373-376

[27] Ouyang Y, Ruan Q, Qian L, Sun L, Zhang Y, et al. Genetic diversity of Daweishan mini, Yunlong short leg and La-bai high leg chickens in Yunnan based on mitochondrial DNA D-loop sequences. Journal of Yunnan Agricultural University. 2017;**32**:70-77

[28] Li Q, Zhang P, Li M, Li L, Hu Y, et al. Genetic diversity and relationship of Dulong chickens using mitochondrial DNA control region. Mitochondrial DNA Part B. 2020;**5**:275-280

[29] Tadano R, Sekino M, Nishibori M, Tsudzuki M. Microsatellite marker analysis for the genetic relationships among Japanese long-tailed chicken breeds. Poultry Science. 2007;**86**:460-469

[30] Luo Q, Wen JK, Wang X, Li ZS. The Breeds of Domestic Animals in Yunnan. Kunming: Yunnan Science and Technology Press; 1987

[31] Yan M-L, Ding S-P, Ye S-H, Wang C-G, He B-L, et al. The complete mitochondrial genome sequence of the Daweishan Mini chicken. Mitochondrial DNA Part A. 2016;**27**:138-139

[32] Weng Z, Xu Y, Li W, Chen J, Zhong M, et al. Genomic Variations and Signatures of Selection in Wuhua Yellow Chicken. PLoS One. 2020;**15**(10):e0241137

[33] Wang Y-M, Khederzadeh S, Li S-R, Otecko NO, Irwin DM, Thakur M, et al. Integrating Genomic and Transcriptomic Data to Reveal Genetic Mechanisms Underlying Piao Chicken Rumpless Trait. Genomics, Proteomics & Bioinformatics. 2021. DOI: 10.1016/j.gpb.2020.06.019

[34] Li L, Li Y. Research on the Status and Problems of Local Chicken Development in Yunnan Province. Rural Economy and Technology, in Chinese. 2017;**28**:64-65

[35] Wen-Ping H. Genetic diversity of Yunnan local chicken breeds and its relationship with origin of Chinese domestic chicken. Biodiversity Science. 1999;7:285

[36] Okamoto S, Inafuku K, Ting Z, Maeda Y, Hou D, et al. Blood protein polymorphisms in native chicken breeds in Yunnan province of China. Animal Science Journal. 2003;**74**:471-476

[37] Yina O, Qian R, Lindong Q, Limin S, Yongyun Z, et al. mtDNA D-loop genetic diversity analysis of Daweishan miniature chicken, Yunlong bantam chicken and Rabo tall chicken. Journal of Yunnan Agricultural University: Natural Science Edition. 2017;**32**:70-77

[38] Wang X, Luo C, Tao Q, Sun L, Li Q, et al. Sequence variation of mtDNA D-loop and origin of Labai High-leg chicken. China Poultry. 2016;**38**:14-18

[39] Li H-F, Chen K-W, Tang Q-P, Shen J-C, Zhang S-J. Analysis of Genetic Diversity of Microsatellite Marker of Chicken Breeds Indigenous to Yunnan Province of China. Journal of Jiangsu Agricultural Sciences. 2006;**38**:33-37, in Chinese

[40] Ye L, Miao Y, Huo J. Genetic diversity of Chahua chicken based on microsatellite markers. Chinese Journal of Zoology-Peking. 2006;**41**:37

[41] Lang Thui Y, Jin Tlong H, Yong Twang M, Sheng Tquan Z, Tao C, et al. Genetic Diversity Analysis of Nixi Chicken Using Microsatellite DNA Markers. Zoological Research. 2006;**27**:68-74

[42] Jia J, Wen S, Xiong B, Ma Z, Miao Y, et al. Genetic diversity of Daweishan Mini chickens based on microsatellite markers. Journal of Yunnan Agricultural University. 2009;**24**:654-660

[43] Qian L, Chen T, Huo J. Microsatellite DNA Variation within the Wuding Chicken Breed. Journal-Yunnan Agricultural University. 2006;**21**:651

[44] Chen T, Y-w M, Huo J-l, Z-y YANG, Z-x LUO, et al. Genetic polymorphism of microsatellite DNA in the population of yanjin black-bone chicken. Journal-Yunnan Agricultural University. 2007;**22**:543

[45] Wang Q, Li D, Guo A, Li M, Li L, et al. Whole-genome resequencing of Dulong Chicken reveal signatures of selection. British Poultry Science. 2020;**61**:624-631

[46] Guo X, Fang Q, Ma C, Zhou B, Wan Y, Jiang R. Whole-genome resequencing of Xishuangbanna fighting chicken to identify signatures of selection. Genetics Selection Evolution. 2016;**48**:62

[47] Liu R, Xing S, Wang J, Zheng M, Cui H, et al. A new chicken 55K SNP genotyping array. BMC Genomics. 2019;**20**:1-12

[48] Wang M-S, Thakur M, Peng M-S, Jiang Y, Frantz LAF, et al. 863 genomes reveal the origin and domestication of chicken. Cell Research. 2020;**30**:693-701

[49] Li D, Che T, Chen B, Tian S, Zhou X, et al. Genomic data for 78 chickens

from 14 populations. Giga Science. 2017;**6**:gix026

[50] Zhang C, Lin D, Wang Y, Peng D, Li H, et al. Widespread introgression in Chinese indigenous chicken breeds from commercial broiler. Evolutionary Applications. 2019;**12**:610-621

[51] Luo W, Luo C, Wang M, Guo L, Chen X, et al. Genome diversity of Chinese indigenous chicken and the selective signatures in Chinese gamecock chicken. Scientific Reports. 2020;**10**:1-14

[52] Tian M, Wang Y, Gu X, Feng C, Fang S, et al. Copy number variants in locally raised Chinese chicken genomes determined using array comparative genomic hybridization. BMC Genomics. 2013;**14**:262

[53] Qu L, Li X, Xu G, Chen K, Yang H, et al. Evaluation of genetic diversity in Chinese indigenous chicken breeds using microsatellite markers. Science in China Series C: Life Sciences. 2006;**49**:332-341

[54] Zhang T, Du W, Lu H, Wang L. Genetic Diversity of Mitochondrial DNA of Chinese Black-bone Chicken. Brazilian Journal of Poultry Science. 2018;**20**:565-572

[55] Miao Y, Peng M-S, Wu G-S, Ouyang Y, Yang Z, et al. Chicken domestication: an updated perspective based on mitochondrial genomes. Heredity. 2013;**110**:277-282

Section 4

Genetic Diversity and Population Structure of Plant Crops

Chapter 5

Perspective Chapter: Nutraceutical Diversity of Eco-Friendly Jute and Allied Fibre (JAF) Crops in Bangladesh

Mohammad Mia Mukul

Abstract

The animal meats contain harmful fat and cholesterol contents but plants are the important sources of secondary metabolites that play important role against diseases, and showed less side effects. Jute, Kenaf and Mesta are self-pollinated annual herbaceous plants used to produce fibre contents. These plants are also used for nutritional purposes. Jute leaves contain vitamins, minerals, energies, macromolecules, phytochemicals, micronutrients, amino acids, anti-oxidants essential to promote human health. Only two species of jute (*C. capsularis* and *C. olitorius*) are commercially cultivated for fibre content. Olitorius leaves are sweet but capsularis leaves are bitter in taste. The young twigs and leaves of olitroius jute are used as delicious vegetable, and some capsularis cultivars released in Bangladesh can be used as leafy vegetables at young stage. C. olitorius is known as Molokhia and used as green leafy vegetable (GLV) in African and Eastern countries as a viscous soup. Hibiscus (Kenaf and Mesta) plants have antioxidants, anticancer, antibacterial and anti-cholesterol qualities. These plants produce multi-coloured lovely flowers used to grace our environment. Kenaf (H. Cannabinus) seeds contain essential amino acids and used as feed meals for the birds in different countries. Mesta or Roselle (H. sabdariffa) leaves and calyces have both vegetable and medicinal values.

Keywords: *Corchorus* spp., cholesterol, *Hibiscus cannabinus*, *H. sabdariffa*, leafy vegetable, metabolites, Molokhia, phytochemicals, vegetable nutrients

1. Introduction

Jute is the English version of the current Bengali word 'Pat or Paat'. Jute is an annual and short day natural fibre plant of the Malvaceae or *Tiliaceae or* currently *Sparminniaceae* family under the *Corchorus* genus [1, 2]. It is a kind of fibre, which is obtained from two commercially cultivated species (*C. capsularis* and *C. olitorius*). It is an important cash crop of Bangladesh. Once it was the golden fibre of Bangladesh not only for the rich golden colour of the fibre but also, metaphorically, for jute's valuable contribution to the national economy of country.

There are over 30 species, which belong to this genus. This plant is grown in the summer season (Kharif-I) for fibre purpose [3]. But it can be grown throughout the year under the temperature above 20°C for vegetable as well as seed production purpose. Nutrition is a basic human need and a prerequisite for a healthy life. Nowadays, people are becoming well aware of the nutritional benefits of fresh vegetables and fruits, and thereby searching for more diversity in their food to get the utmost nutrition. The main focus of consumers is on those foods that are rich in vitamins (A, C, and E), antioxidants, and minerals such as potassium (K), calcium (Ca), and magnesium (Mg).

A diet rich in vegetables and fruits can lower blood pressure, reduce the risk of heart disease and stroke, prevent some types of cancer, lower risk of eye and digestive problems, and have a positive effect upon blood sugar, which can help to keep appetite in check [4]. Eating non-starchy vegetables and fruits such as apples, pears, and green leafy vegetables may even promote weight loss. Their low glycaemic loads prevent blood sugar spikes that can increase hunger. The green leafy vegetables (GLVs) include jute (Molokhia), mesta (Roselle), spinach, lettuce, curly lettuce, chard, purslane, chicory, etc., are rich sources of nutrients, high in dietary fibre, low in lipids, and rich in folate, ascorbic acid, vitamin K, Mg, and K. They also carry plenty of phytochemicals such as β-carotene flavonoids. Jute leaves are important sources of minerals (iron and calcium), vitamins (A, C, and riboflavin), and fibre. Young, fresh leaves contain more vitamin C than mature plants [5]. The green outer leaves of lettuce and cabbage are richer in vitamins, calcium, and iron than white inner leaves. Thinner and greener leaves are more nutritious and usually have lower calories. The GLVs and fresh fruits are gaining a significant place in the food pyramid, being good sources of trace elements and other bioactive compounds. Vegetables are the fresh and edible parts of herbaceous plants extremely valuable for health maintenance and disease prevention. These are mainly appreciated due to their high vitamins, carbohydrates, and especially their mineral contents. The higher the average daily intake of fruits and vegetables, the lower the chances of developing cardiovascular diseases [6]. Most vegetables are naturally low in fat and calories. Diets rich in potassium may help to maintain healthy blood pressure. Vegetables provide nutrients vital for health and maintenance of our body.

2. Genetic diversity and population structure of JAF crops

Population genetics is the branch of genetics that deals with the consequence of Mendelian inheritance or the description of observed or inferred heritable features in population, rather than families through space and time [7]. Population genetics is concerned with genetic differences within and across populations, and the dynamics of how populations evolve as a result of the propagation of genetic mutations occurring within the germlines of individuals [8]. It is the branch of biology that provides the deepest and clearest understanding of how evolutionary change occurs [9]. Population genetics is particularly relevant today in the expanding quest to understand the basis for genetic variation in susceptibility to complex diseases. Population genetics seeks to understand how and why the frequencies of alleles and genotypes change over time within and between populations. The researchers found genetic variations within the same species for certain traits of jute and allied fibre crops. Jute is the second most important global natural fibre crop after cotton. Jute consists of approximately 50–60 species that grow in the tropics, subtropics, and warm temperate regions of the world (mainly in Asia and Africa) [10]. Analyses of the genetic diversity and population structure of the germplasm are useful for improving jute breeding [11]. However, only two cultivated species, *C. capsularis* L. and *C. olitorius* L., are mostly used commercially with an annual global production of 2.65 million tonnes [12].

However, jute possesses narrow genetic diversity since there are limited breeding parent combinations and only a few varieties are cultivated [13].

Therefore, it is important to expand the genetic basis of the breeding populations to promote jute breeding. More than 2000 jute accessions are stored in the National Bast Fiber Germplasm Middle-term Storage, Changsha, China [14]. Moreover, jute accessions derived from the same pedigree may have different names in different places owing to the exchange of germplasm across different countries or regions. The Breeding Division of BJRI has developed 54 varieties of jute and allied fibre (JAF) crops (**Table 1**) through various breeding approaches and selection processes using a large collection of germplasms from different countries of the world [15]. More than 6550 germplasms of JAF crops have been conserved and are maintaining by the Gene Bank under Genetic Resources and Seed (GRS) division of BJRI. The GRS division provides seed materials to the other departments of BJRI and other organizations who are directly involved with JAF crop research activities.

3. Origin, distribution, and morphometric characters of JAF vegetable crops

3.1 Jute (*Corchorus* spp.)

Jute (*Corchorus* spp.) is a diploid (2n = 2x = 14) fibre crop, which ranks second in importance next to cotton as a natural fibre and occupies important place in the

Sl. No.	Varieties & Current name	RY	Pedigree method	RI	Prime purpose	Cultivation status
White Jute/Deshi Pat (*Corchorus capsularis* L.)						
1	Oocarpus	1910	PLS	BJRI	F	N.C.
2	Kakya Bombai	1910	PLS	BJRI	F	N.C.
3	R-85	1916	PLS	BJRI	F	N.C.
4	D-154 (Dacca-154)	1919	PLS	BJRI	F	N.C.
5	D-386	1931	PLS	BJRI	F	N.C.
6	Funduk	1939	PLS	BJRI	F	N.C.
7	C-212	1939	PLS	BJRI	F	N.C.
8	C-13	1941	PLS	BJRI	F	N.C.
9	C-412	1942	PLS	BJRI	F	N.C.
10	C-1	1952	PLS	BJRI	F	N.C.
11	C-2	1952	PLS	BJRI	F	N.C.
12	C-3	1952	PLS	BJRI	F	N.C.
13	C-4 (C-320)	1955	PLS	BJRI	F	N.C.
14	C-5 (C-321)	1955	PLS	BJRI	F	N.C.
15	D-154 (2) [Dacca-154(2)]; Shada Pat	1961	PLS	BJRI	F	N.C.
16	C-6 (C-322)	1967	PLS	BJRI	F	N.C.
17	CVL-1 (Late); Green Pat	1977	PLS	BJRI	F	C

Sl. No.	Varieties & Current name	RY	Pedigree method	RI	Prime purpose	Cultivation status
White Jute/Deshi Pat (*Corchorus capsularis* L.)						
18	CVE-3 (Early); Ashu Pat	1977	PLS	BJRI	F	C
19	CC-45; Joe Pat	1979	PLS	BJRI	F	C
20	BJRI Deshi Pat 5 (BJC-7370); Chaiti	1995	D-154 × CC-45	BJRI	F	C
21	BJRI Deshi Pat 6 (BJC-83); Bijli	1995	CVL-1× Fuleshwari	BJRI	F	C
22	BJRI Deshi Pat 7 (BJC-2142); Basonti	2008	CC-45 × BJC-718	BJRI	F	C
23	BJRI Deshi Pat 8 (BJC-2197); Druti	2013	CC-45 × FDR	BJRI	F	C
24	BJRI Deshi Pat Shak-1 (BJC-390)	2014	Cap Dwarf Red × BINA Pat Shak-1	BJRI	V	C
25	BJRI Deshi Pat 9 (BJC-5003); Basumoti	2017	CVL-1× Acc. 1831	BJRI	F	C
26	BJRI Deshi Pat Shak-2 (Mehra Red)	2020	PLS	BJRI	V	C
27	BJRI Deshi Pat Shak-3 (Mehra Green)	2020	PLS	BJRI	V	C
28	BJRI Deshi Pat 10; Arnob	2021	C-160B × C-164	BJRI	F	C
29	BINA Pat Shak-1 (www.bina.gov.bd)	2003	Mutant (Gamma) of CVL-1	BINA	V	C
Dark Jute/Tossa Pat/Bogi Pat/Misti Pat (*Corchorus olitorius* L.)						
30	Chinsura green (D-38)	1915	PLS	BJRI	F, V	N.C.
31	R-26	1929	PLS	BJRI	F, V	N.C.
32	R-27	1929	PLS	BJRI	F, V	N.C.
33	O-620	1939	PLS	BJRI	F, V	N.C.
34	O-632	1939	PLS	BJRI	F, V	N.C.
35	O-753	1939	PLS	BJRI	F, V	N.C.
36	O-1	1955	PLS	BJRI	F, V	N.C.
37	O-2	1955	PLS	BJRI	F, V	N.C.
38	O-3	1955	PLS	BJRI	F, V	N.C.
39	O-5	1964	PLS	BJRI	F, V	N.C.
40	O-4; Baishakhi Pat	1967	PLS	BJRI	F, V	C
41	O-9897 (Falguni Tossa; Popular variety)	1987	O-5 × BZ-5	BJRI	F, V	C
42	BJRI Tossa Pat 3 (OM-1); Rani Tossa	1995	PLS	BJRI	F, V	C
43	BJRI Tossa Pat 4 (O-72); Sonali Bangla	2002	(O-9897 × O-2012) × O-9897	BJRI	F, V	C
44	BJRI Tossa Pat 5 (O-795); Red Tossa	2008	Uganda Red × O-4	BJRI	F, V	C

Sl. No.	Varieties & Current name	RY	Pedigree method	RI	Prime purpose	Cultivation status
White Jute/Deshi Pat (*Corchorus capsularis* L.)						
45	BJRI Tossa Pat 6 (O-3820); Duronta Tossa	2013	PLS	BJRI	F, V	C
46	BJRI Tossa Pat 7 (MG-1); Basontika	2017	PLS	BJRI	F, V	C
47	BJRI Tossa Pat 8 (Robi-1, Red variety)	2019	TILLING of O-4 and Selection	BJRI	F, V	C
48	JRO-524 (Navin), Exotic	1977	African (cv. Sudan Green) × indigenous (cv. JRO-632)	CRIJAF, India	F, V	C
Kenaf (*Hibiscus cannabinus* L.)						
49	BJRI Kenaf −1 (HC-2); Joli Kenaf	1977	PLS	BJRI	F	C
50	BJRI Kenaf −2 (HC-95)	1995	PLS	BJRI	F	C
51	BJRI Kenaf −3 (HC-3); Bot Kenaf	2010	PLS	BJRI	F	C
52	BJRI Kenaf −4 (KE-3); Red Kenaf	2017	PLS	BJRI	F	C
Mesta (*Hibiscus sabdariffa* L.)						
53	HS-24; Tani Mesta	1977	PLS	BJRI	F	C
54	BJRI Mesta-2 (VM-1); Red Chukur	2010	PLS	BJRI	V	C
55	BJRI Mesta-3 (SAMU'93); Spineless (Smooth) Mesta	2017	PLS	BJRI	F	C
56	BJRI Mesta-4 (VM-2); Green Chukur	2021	PLS	BJRI	V	C

PLS = pure line selection; F = fibre; V = vegetable; C = cultivating; N.C. = not cultivating; RY = release year; RI = releasing Institute; and VM = vegetable mesta.

Table 1.
List of jute and allied fibre crops in Bangladesh with their development method, release year, organization, utilization purpose, and cultivation status [15].

economy Bangladesh and India [16, 17]. White jute (*C. capsularis*) was originated from India or Indo-Burma and south China; and Tossa jute (*C. olitorius*) originated from Africa [18]. *C. olitorius* is known as Molokhia, used as green leafy vegetable, viscous soup in African regions, Middle-Eastern countries, and valued for its nutrient composition [19]. So primary centre of origin of jute is Africa and secondary is Indian subcontinent. The draft genomes of the two cultivated jute species differ with respect to genome sizes. *C. olitorius* has a draft genome size of ~448 Mb, while *C. capsularis* has ~404 Mb [20]. These two species are cultivating in different countries of the world. India, Bangladesh, and China are the world's major producers of jute. Other countries of considerable importance are Brazil, Mexico, China, Venezuela, Egypt, Sudan, Sri Lanka, Middle East, Taiwan, and parts of tropical Africa and Asia. At present, Bangladesh has occupied near about the first position in terms of both production and export of jute and jute products in the world [3]. Information on the genetic diversity within and among closely related crop varieties is essential for crop improvement and to meet the diverse goals such as producing cultivars with increased yield, wider adaptability, desirable quality, pest, and disease resistance [17, 21–25].

There are 80% tossa jute, 7% white jute, and 13% allied fibre crops growing in Bangladesh. Jute is natural fibre and totally biodegradable, cash crop of Bangladesh, called 'golden fibre'. Jute is a rain water dependent crop, and its cultivation requires low amount of fertilizer and pesticide than other crops. The jute plants of 1 m [2] area can absorb 0.23–0.44 mg CO_2. The jute plants of 1 hectare area at 100 days old can consume ~15 tons of CO_2, and release ~11.0 tons of O_2 to the atmosphere and keep the environment clean and rich in O_2. This plant can contribute to save the environment through controlling the greenhouse gas emission as well as rising of environmental temperature. Jute can be used to make paper pulp, and jute sticks are used as fuel, to make charcoal, printer ink, cosmetic's elements, tooth paste, car parts, etc. Jute plant consists of some chemicals, that is, cellulose (70%), hemi-cellulose (20%), lignin (10%); while jute stick contains 30% cellulose and takes about 15 years to produce. Recently, a renowned scientist named Mubarak Ahmad Khan, Scientific Advisor of BJMC and Ex-Director General of BAEC, has developed a new technology from jute fibre. He developed polythene type bag from jute fibre called 'Sonali Bag' (**Figure 1**). The name 'Sonali Bag' was given by Honourable Prime Minister Sheikh Hasina, Government of the People's Republic of Bangladesh in 2015. The Sonali Bag is a biodegradable biopolymer, made of jute cellulose and eco-friendly. It is totally biodegradable, compostable, water soluble; it does not melt in fire like plastic material, which produces ash after burning. The polythene or plastics take long time to decompose in soil, and these are harmful for the environment. Polythene is used in Bangladesh since 1982, and the use of plastic was banned since 2002 [26]. But, plastics are using till now. Jute may be a good source of biopolymer to be used for industrial purposes. The 'Sonali Bag' will be commercialized very soon to be an alarm to plastic bags in Bangladesh. Jute-based other products are personal protection equipment (PPE), face mask, sanitary napkin, etc. Recently, Farhana Sultana, an assistant scientist at ICDDR'B, has won the grand award for her excellent innovation of a machine to make jute cellulose-based disposable sanitary pads for long-term menstrual health and hygiene of women and girls. She has designed and piloted the jute cellulose-based disposable pad project in partnership with Dr. Mubarak Ahmed Khan of BJMC.

Figure 1.
Eco-friendly jute golden bag or Sonali bag made of jute fibre.

3.2 Allied fibre crops

The other two fibre crop species namely Kenaf (*Hibiscus cannabinus* L.) and Mesta (*H. sabdariffa* L.) are also belonging the *Malvaceae* family. Both species were originated in Africa [27]. Kenaf is a tall annual herbaceous woody tropical plant having great potential for fibre, energy, and feedstock [28, 29]. Kenaf is composed of various active components including tannins, saponins, polyphenolics, alkaloids, essential oils, and steroids. Kenaf seed has been the waste part of kenaf plant. It is an annual crop, which is normally cultivated in the tropics and subtropics, where temperatures are greater than 20°C [15]. It is harvested for fibre soon after its flowering. Linoleic, oleic, and palmictic acids were the predominant fatty acids in kenaf seed oil. Under good conditions, Kenaf will grow to a height of 5–6 m in 6–8 months and can yield up to 30–35 t ha^{-1} of dry woody material. It is grown in Bangladesh, Thailand, China, India, Australia, and USA. *Hibiscus sabdariffa* L. (Mesta), also known as Roselle, is an ideal crop for developing countries as it is relatively easy to grow, can be grown as part of multi-cropping systems, and can be used as food and fibre. In Bangladesh, India, China, the Mesta seeds are used for their oil and the plant is used for its medicinal properties, while in West Africa the leaves and powdered seeds are used in meals. Additionally, it is used in the pharmaceutical and food industries. Among the fibre crops except cotton, 7% kenaf and 6% Mesta species are grown in Bangladesh [30]. The popularity of Kenaf is increasing day by day due to its low production costs, fibre and fodder uses, and cultivation in the fellow lands. The fibres of Kenaf & Mesta are coarse type but used for making rope, bags, sacks, etc. in industries.

3.3 Jute and allied fibre crop research in Bangladesh

Jute is one of the mainstays of Bangladesh. It accounts for about 6% of the foreign currency earnings from export. Bangladesh grew 85.76 lakh tonnes of jute in FY2018–19 and 80 lakh tonnes in FY2019–20. Bangladesh currently exports 282 jute and jute-based goods to around 135 countries around the globe. Bangladesh has shut down the Adamjee Jute Mills in 2002. Of the 32 jute mills under the BJMC, 25 remained active after their shutdown last year. These jute mills employ around 2 lakh people. The Bangladesh Jute Research Institute (BJRI) is an oldest mono-crop research institute in the country established in 1951 [30, 31]. Jute Research was first started in Dhaka with the creation of a Fibre Expert's position and assumption of the responsibility by Sir R. S. Finlow in 1904 under the Bengal Department of Agriculture (1904–1939). To intensify research on jute and allied fibres, Jute Agricultural Research Laboratory (JARL) was established by the Indian Central Jute Committee (ICJC) in Dhaka in 1939. BJRI has three main branches, namely 1) Agriculture Research on Jute, 2) Technological Research on Jute, and 3) Jute and Textile Product Development Research. According to Al-Mamun et al. [15], Mukul [32], Haque et al. [33], the Breeding Division of BJRI has developed 54 varieties of jute and allied fibre crops including 28 *C. capsularis*, 18 *C. olitorius*, 4 *H. cannabinus*, and 4 *H. sabdariffa* species (**Table 1**). About 50% of the varieties are in field condition and cultivating by the farmers. White jute is well adapted here than tossa jute but fibre quality and market demand are high for tossa jute. So that farmers cultivate more tossa jute than white jute in Bangladesh. Kenaf crop is used as a good source of pulp for paper production as well as for industrial uses. Kenaf seeds are good source of essential amino acids and other nutrients, where Mesta crops are less demanded by the farmers due to less utilization in industries, but leaves and fruit calyces are used as vegetables and used to make various food staffs. Kenaf and Mesta need very less cost of production. These plants can be grow well in fellow, less fertile, uncultivated lands with low organic matter content [34].

4. Jute, Kenaf, and Mesta are sources of green leafy vegetables and essential nutrients

In Bangladesh, the young-fresh leaves and shoot tips of *C. olitorius* varieties (**Figure 2**) are used as leafy or green leafy vegetables (GLVs) due to its sweet taste [11]. It is also used as Jew's Mallow Plant in Africa and as Sorel leaves in Asia [35]. Molokhia (*C. olitorius*) is a popular summer vegetable dish due to its special delicious taste; it is consumed fresh or dried in vegetable soup. White jute (*C. capsularis)* leaves and twigs (**Figure 3**) are bitter in taste but some varieties (**Figures 4–7**) have been

Figure 2.
Young leaves and twigs of C. olitorius *jute (O-9897) used a vegetable.*

Figure 3.
Young leaves and twigs of C. olitorius *jute (CVE-3) grown mainly for fibre and slightly used a vegetables.*

Figure 4.
Young leaves and twigs of BJRI Deshi Pat Shak 1 (BJC-390) used as vegetable.

Figure 5.
BJRI Deshi Pat Shak 2 (Mehra red) cultivated in field.

developed as vegetable type with no bitterness, which is consuming in a large volume by the people. The leaves of *H. cannabinus* (**Figure 14**) is largely consumed by the domestic animals (cow, buffalo, goat, etc.) and seeds (**Figure 15**) are used to prepare feed meal of layar & broiler chicken, and to extract oil content with good amount of amino acids and other nutrients. The calyces and young leaves (**Figure 16**) of *H. sabdariffa* are used as fresh vegetables; seeds (**Figure 17**) are used for feed meals and oil purposes; and the fruit calyces are more popular for making drinks, ice creams, jellies, jams, pickles, cuisine, etc. (**Figure 18**).

Figure 6.
BJRI Deshi Pat Shak 3 (Mehra green) cultivated in field.

Figure 7.
BINA Deshi Pat Shak 1 cultivated in field.

5. Biochemical basis and nutritional profile of jute leaves

There are two types of vegetable jute in our country, such as Desi or White Jute and Tossa or Bogi or Dark jute. So far, it has been found that vegetable jute (Pat shak) contains about 17 active nutrient elements such as meat, oil, sugary fibre, ash, calcium, potassium, iron, sodium, phosphorus, beta carotene, thiamine, riboflavin, niacin, ascorbic acid, vitamin A, and water (**Table 2**) [36]. Deshi Pat Shak has a bitter taste due to its presence of capsin. However, since tossa jute is not bitter, it is used in making various types of soups. The nutritional value of jute leaves is astonishing. However, due to its availability and cheapness, people are not aware of its nutritional value. At present, different species of vegetables are available in the market in winter in Bangladesh, such as Puishak (*Basella alba*), Spinach (*Spinacia oleracea*), Amaranth (*Amaranthus tricolor*), kachushak (*Colocasia esculenta*), Radish Spinach (*Raphanus sativus*). Of all these vegetables, spinach is the most well-known, nutritious, valuable,

Nutrients	1 cup (28 grams), raw jute leaves	1 cup (87 grams), cooked jute leaves
Energy (Calories)	10.0	32.0
Protein	1.0 g	3.0 g
Fat	0.07 g	0.17 g
Carbohydrate	2.0 g	6.0 g
Fibre	0.0 g	2.0 g
Minerals		
Calcium	4% of the daily value (DV)	14% of the DV
Magnesium	4.0% of the DV	13% of the DV
Potassium	3.0% of the DV	10% of the DV
Iron	7.0% of the DV	15% of the DV
Vitamins		
Vitamin A	9.0% of the DV	25% of the DV
Riboflavin	12% of the DV	13% of the DV
Folate	9.0% of the DV	23% of the DV
Vitamin C	12% of the DV	32% of the DV

Table 2.
Proximate composition of jute leaves [36].

and favourite vegetable. Comparing the nutritional value of jute with this spinach, the importance of nutritional value of jute will be easily understood. The nutritional composition of jute plant depends on plant leaves maturity. It was observed in an earlier research report, high-moisture content of jute leaves was decreased from 15th days after seed sowing to 33th days old. Leaf maturity had significant ($p < 0.05$) effect on the moisture of jute leaves. The ash content of jute leaves decreases at 15th to 33th days old and then increases with increasing of plant age up to maturity. In biota, ash content is an index of mineral contents, which is relatively high in jute leaf comparing to the values found in *Hibiscus esculentus* (8.00% DW) reported by Akindahunsi and Salawu [37]. The ash content within the range of 16.30–17.31% reported for some vegetables by Dairo and Adanlawo [38]. The crude protein content increased from 15th DASS (19.86%) till 24th DASS (26.46%) and reduced to 16.65% till final harvesting at 33th DASS. It indicates that jute leaf is advantageous as a rich source of vegetable protein over some vegetables such as raw Cocoyam leaf (3.4%), cooked Cocoyam leaf (2.1%), Amaranthus (6.1%), and *Moringa oleifera* (4.2%) as reported by Adepoju et al. [39]. According to Ali et al. [36], the maturity of jute leaf does not affect its fat content ranging from 1.54 to 2.95%; it revealed that jute plant is lower in fat content like *Crassocephalum crepidioides* (12.45%) and *Senecio biafrae* (14.21%) reported by Dairo and Adanlawo [38], but higher than value reported for *Brassica oleracea* (0.26%) by Emebu and Anyika [40]. The fibre content of jute leaves increases with the increase of leaf maturity [36]. However, jute leaves showed higher fibre content (12.04%) than *Amaranthus hybridus* (8.61%) reported by Akubugwo et al. [41]. Foods containing fibres remove potential carcinogens from the body, cleanse the digestive tract, and prevent excess cholesterol absorption. Fibre also adds bulk to the food and prevents the intake of excess starchy food and may therefore guard against metabolic conditions such as hypercholesterolemia and diabetes mellitus [42].

6. Comparison among the nutritional values of jute vegetable, spinach, and amaranth (per 100 g)

The nutritional value of jute leaves is very surprising. Many vegetables are available in the market, such as Puishak, Palangshak, Datashak, Kachushak, Mulashak. Of all these vegetables, spinach is the most well-known, nutritious, valuable, and beloved vegetable. A comparative analysis of the nutritional value of jute with spinach and amaranth will make clear the importance of nutritional value of jute (**Table 3**) [43]. Jute leaves are widely used in rural areas of Bangladesh. But if the leaves are needed in other seasons, then the leaves can be dried and powdered and stored carefully.

Nutrient elements and unit	Jute vegetable	Spinach	Amaranth
Energy (kcal)	73.00	23.00	18.00
Carbohydrates (g)	52.52	3.60	0.30
Dietary/crude fibre (g)	6.80	2.20	2.60
Crude protein (g)	3.77	3.00	n.d.
Lipid (g)	0.60	0.20	n.d.
Calcium (mg)	298.00	55.00	270 g
Iron (mg)	11.00	3.90	3.00
Carotene (mg)	1.92	1.56	1.73
Vitamin B-1 (mg) Thiamine	15.00	0.13	0.07
Vitamin B-2 (mg) Riboflavin	28.00	0.23	0.22
Vitamin-C (mg)	64.00	65.00	42.00

n.d. = not detected.

Table 3.
Comparison jute vegetable, spinach, and amaranth for nutritional compositions [43, 44].

7. Deshi pat or white jute (*Corchorus capsularis* L.) as vegetables

The white jute is cultivated mainly for fibre along with vegetable purpose. It is grown in Kharif-I season for fibre production and year-round cultivation of vegetable species for shak or vegetable production in Bangladesh. The *C. capsularis* plant generally becomes 2.50–3.80 m in height. The stems are angular in shape, green, or greenish-red, and the growth of the periderm in the stems varies at different stages of maturation. Leaves are small, thin, rough, and bitter in taste. However, BJRI Desi Pat Shak-1, 2 and 3 are not bitter. The flowers are small; the sepals and ovaries are pale yellowish and round. White jute fruits are globose type and five chambered, and seeds are small and triangular in size (**Figure 8**), which are green to grey in colour, and 1000 seeds weight is 2.0 g. The seeds are slightly larger and the colour is usually brown (**Figure 8**). However, the seeds of BJRI Desi Pat 7 (BJC-2142), the only white jute variety invented by BJRI, are bluish. The weight of 1000 seeds is about 2.90–3.30 g. Fibre colour is white or snow white (**Figure 8**). The amount of fibre cuttings is 15–20%. White jute is tolerant to many biotic (diseases, pests, insects) and abiotic (floods, drought, salinity, waterlogging) stresses [23, 45]. High-to-medium high land, loam, sandy loamy, and clay loam soils with drainage facilities and organic

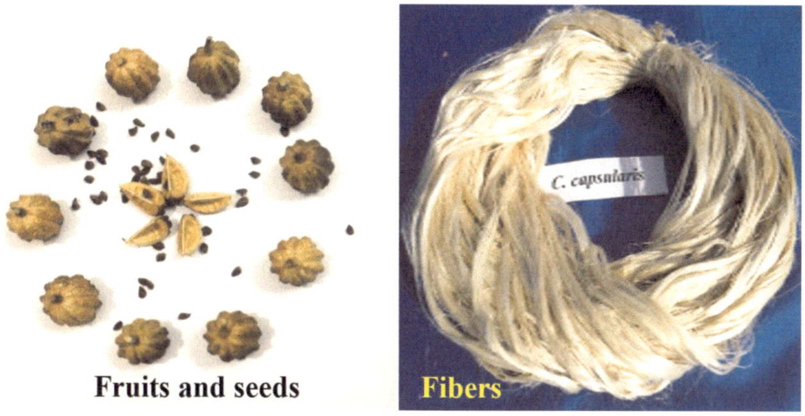

Figure 8.
*Globose-type fruits, triangular-shaped seeds, and snow white fibres of Deshi/White jute (*C. capsularis*).*

matter contents are suitable for white jute cultivation. White jute seeds can be sown through broadcasting or in rows. Line sowing method is easy for intercultural operations, low operational costs, and higher fibre yield and qualities. The BJRI has developed three white jute vegetable varieties and the BINA developed an white jute vegetable variety (**Figure 7**), which are good sources of vitamins, minerals, protein, and moisture content (**Table 4**) [46–48].

More amounts of seeds are required for broadcasting than line sowing. For broadcasting, 30 g seeds are required per decimal or 7.50 kg seeds per hectare of land. On the other hand, 25 g seeds are required per decimal or 6.25 kg seeds per hectare of land for line sowing method. The distance from row to row is 30 cm, plant to plant is 6–7 cm, and plant population: 30 plants m^{-2} or 3.0–3.5 lac plants ha^{-1} of land should be maintained to get higher fibre yield and qualities. Since this species was originated in Indo-Burma sub-continent, it is very adaptive to the soil and environment of

Analyses components	Advanced line	BINA Pat	BJRI Deshi Pat		
	Capsularis dwarf red	Shak-1	Shak-1	Shak-2	Shak-3
Protein (%)	2.37	22.83	22.66	18.45	18.45
Ash (%)	1.68	5.69	8.58	7.77	7.770
β-carotene (μg g^{-1})	145.85	95.59	115.8	125.4	125.4
Vitamin-C (mg per 100 g)	0.0034	63.85	67.73	74.53	74.53
Moisture (%)	78.61	80.63	84.08	85.04	85.04
Calcium (%)	n.d.	1.91	1.62	2.36	1.93
Potassium (%)	n.d.	1.39	1.75	1.61	1.67
Sodium (%)	n.d.	0.091	0.111	0.125	0.118
Phosphorus (%)	n.d.	0.636	0.654	0.62	0.588
Iron (ppm)	n.d.	1344.0	993.0	971.0	610.0

n.d. = not detected.

Table 4.
Biochemical analyses of white jute vegetable type varieties [46–48].

Bangladesh, India, and other nearest countries. Due to the bitterness of plants or leaves, the antioxidant contents are quietly more in white jute. All the varieties of white jute are not eaten as vegetable. Bangladesh Jute Research Institute (BJRI) has developed 29 *capsularis* varieties where there are three varieties for vegetable purpose and 26 varieties for fibre purpose. Among 29 white jute varieties, 13 varieties (Table: 1; SL: 17–29) including 10 fibrous and 03 vegetable types are cultivating now in farmers' fields. Another *capsularis* vegetable variety named BINA Pat Shak-1 (**Figure** 7) has been developed by the Bangladesh Institute of Nuclear Agriculture (BINA) through physical mutation using gamma radiation on the CVL-1 (BJRI Deshi Pat variety) during 2003. The vegetable species of white jute are rich sources of proteins, carbohydrates, energies, vitamins, and minerals. The nutritional status of white jute vegetable species in Bangladesh are summarised here (**Table 4**) [46–48].

8. Tossa pat or Bogi pat or dark jute *(Corchorus olitorius* L.*)* as vegetables

The dark jute is not native to Bangladesh, and it was originated from African regions. It is cultivated for fibre production in Kharif-I season, and it can be used as vegetable. It is known as Molokhia or Jew's Mellow Plant or Nalta jute used as vegetable as well as to make viscous soup in African regions [43]. The plants become 2.8–4.8 m tall. Its stems are tubular, green, or light red to dark red in colour depending on sunlight for red pigmentation, missing lenticels present in periderm. Tossa jute leaves are dark green, thick and smooth, and not bitter in taste. The flowers are relatively large, the petioles are green, the petals are yellow, and the ovaries are oblong. The fruits are long in shape and divided by seeds in four rows. Each row contains 25–40 seeds. Each fruit contains about 130–200 seeds. Tossa jute fruits are capsule or pod type, five chambered, seeds are small and triangular in size, green to grey or bluish green in colour, and weight of 1000 seeds is 1.8 g (**Figure 9**). Seed colour is usually

Figure 9.
*Capsule or pod-type fruits, triangular-shaped seeds, and golden fibres of tossa jute (*C. olitorius*).*

bluish or bluish green (**Figure 9**), but OM-1 variety is brown in colour [49]. Fibre colour is from golden to light butter-white (**Figure 9**), which is very bright and clean flawless fibre. The amount of cuttings is less than that of white jute. They cannot tolerate waterlogging, and photosensitive, that is, premature flowering occurs due to early sowing before 15 March leading to determinate the vegetative growth or fibre production. High-to-medium high land, loam to sandy loamy fertile soils with drainage facilities, and organic matter contents are suitable for tossa jute cultivation. Seeds can be sown through broadcasting or in rows. Line sowing method is easy for intercultural operations, low operational costs, and higher fibre yield and qualities. More amounts of seeds are required for broadcasting than line sowing. For broadcasting, 25 g seeds are required per decimal or 6.50 kg seeds per hectare of land. On the other hand, 20 g seeds are required per decimal or 5.0 kg seeds per hectare of land for line sowing method. The distance from row to row is 30 cm, plant to plant is 6–7 cm, and plant population: 30 plants m^{-2} or 3.0–3.5 lac plants ha^{-1} of land should be maintained to get higher fibre yield and qualities. Since this species was originated in Africa, it cannot tolerate biotic and abiotic stresses in the soil and environment of Bangladesh. The breeders of BJRI are trying and already developed/identified some biotic stress tolerant germplasms, which will be used in hybridization systems and stress-tolerant tossa jute varieties will be developed. They are also taking initiatives to develop hybrid jute varieties using cytoplasmic male sterility (CMS) systems in Bangladesh. Bangladesh Jute Research Institute (BJRI) has developed 18 tossa jute varieties for fibre purpose (Table: 1; S.L. No.: 30-47). An exotic tossa jute variety named JRO-524 (Navin) was introduced in Bangladesh from India. Among the 18 tossa jute varieties of BJRI, 8 varieties (Table: 1; S.L. No.: 40-47) are cultivating along with the JRO-524 by the farmers. All tossa varieties can also be eaten as vegetable at young stage. JRO-524 was developed by Central Research Institute for Jute and Allied Fibres (CRIJAF), India, through crossing bet African cv. Sudan green and indigenous cv. JRO-632 in 1977 [50, 51]. The tossa jute species are rich sources of proteins, carbohydrates, energies, vitamins, and minerals. Various nutritional uses of white jute vegetable species in Bangladesh are summarised here.

9. Preparation of various food items from jute vegetables

9.1 Fresh jute vegetables

Jute is one of the favourite vegetables of many people in Bangladesh. It is also full of various nutrients. Any dish of jute spinach with hot rice is delicious to eat. The tip of the young jute plants should be cut and washed well. The tip of young jute plants at the age of 20–40 days old should be cut off and washed well in clean water followed by draining of water completely. Then, take the jute leaves in a pan and add salt to it. After boiling for 12–15 minutes, drain the water with the slippery juice of the jute leaves. Otherwise, it will not be delicious to eat. Then, 2–3 dried chillies, onion, garlic should be fried with mustard or soybean oil in a pan till it turns brown, and then add boiled jute leaves and fry it. If you need to give salt again, you have to give it. After that, cover the fried jute vegetable for 5 minutes and this cooked jute vegetable (**Figure 10**) is ready to eat with hot rice, pulses, and mash.

Figure 10.
Cooked vegetable of fresh jute.

9.2 Fresh jute vegetables and pulses

Jute vegetables and pulses are also a food known as 'Shukto' that brings water to the tongue (**Figure 11**). The tip of the young jute should be cut and washed well. Now fry the black cumin seeds, dried chillies, and pulses in a pan with oil. After that, fry in a pan with jute and some amount of raw chillies and cover 2 minutes. Then, open the lid, and shake it with salt and turmeric and cover again. When the water of vegetables is dry and soft, it should be mixed with boiled pulses. When mixed with pulses, it tastes very good when eaten with hot rice.

9.3 Fresh jute vegetable's Sholka

This is one type of food item cooked with jute vegetable, which is known as Sholka or Paelka (local) (**Figure 12d**) and popular at northern region Rangpur district of Bangladesh. However, it is not cooked in all areas of Rangpur. The young jute leaves should be cut into small pieces (**Figure 12a–c**) and add some other vegetables such as Gourd leaves, Pumpkin leaves, *Basella* leaves, *Colocasia* leaves, Moringa leaves to get

Figure 11.
Cooked jute vegetables with pulses known as Shukto in Bangladesh.

Figure 12.
(a) Collection of young leaves and twigs of jute, (b) leaves are chopped, (c) chopped leaves are crushed, and (d) finally cooked jute vegetables known as Sholka at Rangpur of Bangladesh.

rid of its bitterness. One of its ingredients is baking soda. A pinch of baking soda is given to lubricate the jute leaves. Now add a little ginger powder and stir for 10 to 15 minutes on low heat. Then, it should be covered and kept for some time. After that, it tastes great when served with food. Also, jackfruit seeds would be added with this item at cooking time, and its taste will increase a lot more.

9.4 Dried jute vegetables

Dark green dried jute leaves are very popular among Japanese users (**Figure 13b**). A preliminary test showed that if the jute leaves were dried in an electric oven for 8 hours at a temperature of 32°C, the colour of the leaves remained green without loss (**Figure 13c**). In this condition, the amount of water in leaves is about 8%. Such dried leaves can be powdered to the size of 16 mash-sized filters (**Figure 13d**). However, in general, jute leaves can be dried at room temperature and powdered and used as a vegetable or filling material. In addition, the green leaves of jute can be chopped (**Figure 13a**), dried in humid air (**Figure 13c**) and fried using garlic, onion, mustard soybean oil, dried chillies, and eaten very tasty with rice and pulses. At present, it is a popular food item in Mymensingh district of Bangladesh.

10. Kenaf (*Hibiscus cannabinus* L.) and its uses

Kenaf is an annual bast fibre crop originated from Africa, and used as a source of fibre to many industries and medicinal constituents [52]. This crop is diploid (2n = 2x = 36) in nature, and belongs to *Malvaceae* family and section *Furcaria* [53] along with hibiscus (*Hibiscus hibiscum* L.), Hollyhock (*Althaea rosea*), Cotton (*Gossypium hirsutum* L.), and Okra (*Hibiscus esculentus*), and is grown commercially in different countries including Bangladesh, India, Pakistan, China, Malaysia [54]. It is

Figure 13.
(a) Chopping and crushing of jute fleshy leaves, (b) packaging of dried jute leaves in green condition, (c) dried and crushed jute leaves in green condition, and (d) grinded powder of jute leaves in green condition for vegetable or drinks or medicinal purposes.

an alternative crop that may be a feasible source of eco-friendly cellulose. The seed capsule (fruit) that is about 1.9–2.5 cm long and 1.3–1.9 cm in diameter and hairy and contains five segments that are many-seeded (20–26 seeds) [55]. Kenaf leaves and seeds (**Figures 14** and **15**) are also using in traditional medicine in India and Africa for the treatment of various disease conditions. Kenaf plants or leaves (**Figure 14**) are used as feed meals for the domestic animal (cow) in Bangladesh. The seeds are brown, glabrous, wedge-shaped, 6 mm long, 4 mm wide, and their weight is about 35,000 to 40,000 seeds kg^{-1} corresponding to 25–29 g thousand grain weight (**Figure 15**). Kenaf fibres (**Figure 15**) are commonly used for paper pulp and cordage, but it is also a promising lignocellulosic feedstock for bioenergy production. The kenaf seed oil can be used for cooking and in different industrial applications. It is an important bast fibre crop and cultivated for fibre purpose. It contains long and short fibres in its stem fractions, bark and core, respectively, which has numerous industrial applications *viz.*

Figure 14.
H. cannabinus *varieties: Young leaves, twigs of HC-95, and flower, fruits, and leaves of HC-2.*

Figure 15.
Seeds and fibres Kenaf (HC-95).

paper and pulp, fabrics, textiles, biocomposites, insulation mats, absorption materials, animal bedding [56]. Kenaf plants possess a wider range of adaptation to environments, climates, soils, and are rich sources of cellulose compared with any of other fibre plant in profitable manufacture industry [54]. At the beginning of the eighteenth century, kenaf was introduced into southern Asia and was first cultivated and commercially utilized in India. The knowledge of how kenaf was introduced in India is limited but it is known that it came from Africa. The cultural interaction between ancient Egypt and the Indus may have played an important role for kenaf's dissemination from Africa to India, from where kenaf cultivation was expanded to other Asian countries. In the beginning of 1900, kenaf was disseminated into mainland China from Taiwan. Currently, many countries pay more attention to kenaf research and cultivation because of its high biological efficiency and wide ecological adaptability. Kenaf is more commonly called 'the future crop'. Nowadays, kenaf is commercially cultivated in more than 20 countries, particularly in China, India, Thailand, and Vietnam [57]. Kenaf leaves extract (KLE) can be used as functional ingredient in cosmetic formulations. It introduces new possible application for kenaf leaves as high value-added ingredients with skincare properties for the cosmetic industry, namely antioxidant, anti-aging, and anti-melanogenic activities. Kenaf can be used for food-fodder-forage. Kenaf seed is a valuable component of kenaf plant. For several years, it has been primarily used as cordage crop and secondarily as a livestock feed. Consumers are becoming more interested in naturally healthy plant-based food products. Kenaf seed, the future crop with a rich source of essential nutrients [58, 59] (**Table 5**) and an excellent source of phyto-compounds, might serve suitable roles in the production of value-added plant-based foods. Kenaf seeds have many nutritional functions and could be used as a valuable natural source of ingredient for the production of functional foods. Kenaf seeds are an important source of dietary fibres, oil, and proteins. The oil has been suggested to be used as a new source of functional edible oil with high antioxidant activity [60] and anticancer properties [61].

Several researchers have reported the value of kenaf seed-derived products like kenaf seed flour (KSF), defatted kenaf seed meal (DKSM), and kenaf seed protein concentrates (KSPC); they have health-supporting activities [62]. The potential of kenaf seed as a source of functional edible oil seems to be excellent [60, 63] as it contains alpha-linolenic acid, an essential omega-3 fatty acid with anti-inflammatory and

Essential amino acids	Whole kenaf seed (mg/100 g)	Defatted kenaf seed (g/100 g)	Protein concentrate (g/100 g)
Histidine	8.92–15.70	5.24–12.50	22.20–80.90
Isoleucine	13.02–16.02	0.26–2.44	0.97–3.15
Leucine	11.38–17.74	1.72–2.57	1.91–2.73
Lysine	2.34–3.86	3.83–4.58	4.59–5.27
Methionine	2.58–4.28	0.11–0.43	0.37–4.79
Phenylalanine	31.79–38.20	0.53–1.45	1.08–3.04
Threonine	20.08–28.00	1.21–3.68	0.95–4.50
Valine	38.13–50.03	0.80–1.36	1.10–1.63

Table 5.
Essential amino acid content of whole, defatted, and protein concentrate of different kenaf seed cultivars [27, 59].

Compositions	(g/100 g)	Compositions	(g/100 g)	Reference
Oil content	23.70%%	Oleic acid	29.2%	[67]
Total phospholipids	6.00%	Linoleic acid	45.9%	
Sterol	0.9%	Palmitoleic	1.6%	
Palmitic acid	20.1%	Linolenic acid	0.7%	
Sphingomyelin	4.42%	Stearic acid	3.5%	
Phosphatidyl choline	21.9%	Phosphatidyl thanolamine	12.8%	
Phosphatidyl inositol	2.7%	Phosphatidyl serine	2.9%	
Phosphatidyl glycerol	8.9%	Lysophosphatidyl choline	5.3%	
Cardiolipin	3.6%	Phosphatidic acid	4.9%	
Stigmasterol	6.07%	β-sitosterol	72.3%	
Campesterol	9.9%			
Calories	427 cal/100 g	Polyunsaturated fat (cis)	6.47 g/100 g	[USDA]
Carbohydrate, total	34.5%	Monounsaturated fat (cis)	5.07 g/100 g	
Ash	5.91%	Iron	12.8 mg/100 g	
Calcium	295 mg/100 g	Moisture	7.69%	
Cholesterol	<1.0 mg/100 g	Potassium	1.290 g/100 g	
Dietary fibre, total	27.6 g/100 g	Protein	35.6 g/100 g	
Fat (total triglycerides)	16.3 g/100 g	Sodium	30.9 mg/100 g	
Saturated fat	4.77 g/100 g	Sugars, total	3.09 g/100 g	
Trans fat	<0.10 mg/100 g			

USDA = United States Department of Agriculture.

Table 6.
Biochemical compositions of kenaf seeds [67].

antithrombotic activities, and also chemopreventive activity [64]. The relatively high oil composition like that of cottonseed oil, and appreciable quantities of phospholipids and phytosterols suggest that kenaf oil can be used for culinary purposes and the seeds could serve feed and food purposes [58, 65]. Researchers have shown interest in isolating the bioactive components (phytosterol) of kenaf seed oil for the production of healthy and nutritious foods. The phenolic and flavonoid compounds present in kenaf seed have been reported as potential inhibitors of angiotensin I-converting enzyme and in the peroxidation of lipids [66]. The kenaf seeds contain 19.84% oil and 13.5% protein content [30]. Kenaf seed oil contains alpha-linolenic acid (ALA), the essential omega 3 fatty acid that is metabolized to Eicosapentaenoic acid, a precursor of eicosanoids with anti-inflammatory and antithrombotic activities. Mohamed et al. [67] evaluated the oil, fatty acid, phospholipid, and sterol content of nine American kenaf genotypes. They reported the biochemical compositions of kenaf seeds (**Table 6**).

11. Mesta (*Hibiscus sabdariffa* L.) and its uses

Hibiscus sabdariffa commonly named as 'Red Sorrel' or 'Roselle' is a member of Malvaceae family used for fibre production. There is a big argument about the origin of Roselle among different scholars. *Hibiscus sabdariffa* L. (Roselle) is an annual plant (**Figure 16**) that has traditionally been used in human and animal diet as food, drinks, and medicine (**Figure 18**). Roselle is a native plant of West Africa and from there, it was carried to other parts of the world such as Asia and America, whereas in others opinion, Roselle was originated from India and Saudi Arabia [68]. *H. sabdariffa* can be used as medicine for the treatment of high blood pressure, liver diseases, and fevers [69, 70]. In large amounts, hibiscus tea acts as a mild laxative. In Iran, it is a traditional treatment for high blood pressure, which is the focus of several studies, as is cholesterol reduction. Roselle is rich in organic acids including citric, malic, tartaric, and allo-hydroxycitric acids. The plant is also known for its β-carotene, vitamin C, protein, and total sugar. Roselle, having various medically important compounds called photochemical, is well known for its nutritional and medicinal properties. Many parts of Roselle including leaves and tender shoots, stems, calyces, fruits, seeds, and roots are used in various

Figure 16.
Leaves (c), twigs, fruits with calyces (a, b) of H. sabdariffa *(VM-1: Red chukur and VM-2: Green chukur).*

Figure 17.
Seeds and fibres of Mesta (VM-1).

foods as well as in herbal medicine as a potential non-pharmacological treatment. They are sources of anthocyanin, flavonoids, polyphenols, organic acids, and fibre, which are single-stand products, and beverages and medicinal products are obtained from them. The plant also acts as an antioxidant and used in obesity management. Its extracts showed antibacterial, antioxidant, nephro- and hepatoprotective, renal/diuretic effect, effects on lipid metabolism (anticholesterol), and antidiabetic and antihypertensive effects among others. It is an important cash crop grown in almost all warm countries such as Bangladesh, India, south of China, Saudi Arabia, Malaysia, Indonesia, Thailand, Philippines, West Africa, Vietnam, Sudan, Egypt, and Mexico [71]. Mesta seeds (**Figure 16**) can be used for feed meals and oil purposes. Its fibres (**Figure 17**) are very strong but coarse type is used for making sacks, ropes, and other goods. Some products of Roselle include sauces, vegetable salads, fruit salads, creams, perfumes, marmalade, seasoning products, fibres, spices, sauces, and vegetable oils. The red-coloured calyces of Roselle can be used as vegetable, to make jam, jelly, pickles and colouring raw materials of ice cream; dried calyces are used to prepare drinks (**Figure 18**). The brilliant red colour and unique flavour make it a valuable food product. It is also rich in bioactive compounds such as anthocyanins and other flavonoids, organic acids, and polysaccharides that are responsible for its antioxidant, antibacterial, anti-inflammatory, hepatoprotective, and anticholesterol activities. Different extracts from Roselle play a crucial role in treating different medical problems including many cardiovascular disorders, helmenthic disease, and cancer. The calyces of Roselle are rich in anthocyanin, ascorbic acid, and other phenolic compounds. Roselle plant is an important source of energy, protein, vitamins, and minerals (**Table 7**) [31]. It is water soluble with brilliant and attractive red colour and with sour and agreeable acidic taste, which aid digestion. Roselle has been used by people for preparing soft drinks and in traditional medicine. It has been observed that its components, such as vitamins (C and E), polyphenols acids, and flavonoids, mainly anthocyanin, have functional properties (**Table 8**) [72]. They contribute benefit to health as a good source of antioxidants as well as a natural food colourant. The other health benefits of this plant include diuretic

Figure 18.
Preparation of different food items from the Roselle or Mesta calyces (VM-1: Red chukur): (1) fruits with calyces, (2) fresh calyces, (3) dried calyces, (4) Roselle drink (tea), (5) drying & crushing of calyces, (6) prepared jelly, (7) ice creams, (8) cold drinks, (9–10) fresh calyces and boiling, (11) prepared Roselle jam, and (12) pickle.

and choloratic properties, intestinal antiseptic, and mild laxative actions. It also used in treating heart and nerve disorder, high blood pressure, and calcified arteries. Due to perceived safety and physiological advantage of the natural colourants over synthetic ones, interest is being geared into the search of new natural colourants and the verification of the safety of existing ones.

Elements	Roselle calyces		
	Green	Red	Dark red
Crude protein (%)	17.9	17.4	8.6
Ether extract (%)	3.2	2.1	2.9
Crude fibre (%)	11.2	8.5	9.8
Ash (%)	6.6	6.5	6.8
Ascorbic acid (mg/100 g)	86.5	63.5	54.8
Moisture (FW)%	88.3	86.5	85.3
Calcium (mg/100 g)	1209.0	1583.0	1602.0
Magnesium (mg/100 g)	235.0	316.0	340.0
Potassium (mg/100 g)	1850.0	2060.0	2320.0
Sodium (mg/100 g)	9.5	5.5	6.5
Iron (mg/100 g)	32.8	37.8	34.6
Zinc (mg/100 g)	5.8	6.5	6.3

Table 7.
Chemical composition of Roselle calyces [72].

Nutrients	Per 100 g raw	Biochemical values of different parts		
		Seeds	Leaves	Calyxes
Energy	49 kcal	n.d.	n.d.	n.d.
Protein	0.96 g	28.90 g	3.50 g	2.00 g
Carbohydrate	11.31 g	25.50 g	8.70g	10.20 g
Fat	0.64 g	21.40 g	0.30 g	0.10 g
Vitamins				
Vitamin A equiv.	14.0 μg	n.d.	1000 (I.E.)	n.d.
Thiamine (B_1)	0.011 mg	0.10 mg	0.20 mg	0.05 mg
Riboflavin (B_2)	0.028 mg	0.15 mg	0.40 mg	0.07 mg
Niacin (B_3)	0.31 mg	1.50 mg	1.40 mg	0.06 mg
Vitamin-C	12.00 mg	9.00 mg	2.30 mg	17.00 mg
Minerals				
Calcium	215.0 mg	350.0 mg	240.0 mg	150 mg
Iron	1.48 mg	9.0 mg	5.0 mg	3.0 mg
Magnesium	51.00 mg	n.d.	n.d.	n.d.
Phosphorus	37.00 mg	n.d.	n.d.	n.d.
Potassium	208.00 mg	n.d.	n.d.	n.d.
Sodium	6.00 mg	n.d.	n.d.	n.d.

n.d. = not detected.

Table 8.
Nutritional values of Roselle plant [31].

12. Preparation of Roselle drink

The Roselle or Chukur's drink can be prepared through the following steps:

- Fleshy fruits of red (**Figure 18**) or green coloured calyces of Roselle should be collected and washed well in clean water followed by air drying or oven drying for 3 days at 70°C temperature;
- Calyces should be peel off and stored in airtight container.
- Simply, 2.0 g of dried calyx should be crushed into small pieces using a wooden roller to make the Roselle drink (**Figure 18**);
- Then those should be put into a tea bag or net; the bag should be steeped into boiled water in a cup; sugar could be added along with the inclusion of lemon juice to make flavour if desired;
- The Roselle drink is ready to enjoy. The dried calyx could be refrigerated and used to make Roselle iced drink as and when necessary (**Figure 18**).

13. Medicinal/nutritional as well as pharmacological properties of jute and allied fibre crops

The medicinal properties of jute have been mentioned in the medical textbook 'Charak Sanghita' that it is effective in destroying leprosy, relieving urinary tract infections and relieving arthritis pain. In Ayurvedic medical science, the juice of tita (bitter) jute or white jute leaves is an infallible medicine for diarrhoea, fever, and acidity. It increases appetite and digestion and it is very useful as a constipation cleanser. Again, a mixture of sweet or tossa jute leaf juice and turmeric powder cures complicated bloody diarrhoea. Drinking water by soaking the leaves of tossa jute increases urination, eliminates inflammation of the bladder, and increases the strength of the body. Compared with other vegetables, jute leaves are rich in energy, calcium, iron, carotene (vitamin A), and vitamin C. Therefore, the importance of using jute leaves as a vegetable and as a blood cleanser for various problems of the body as medicine is immense. Jute juice or soup can be used to treat fever, chronic cysts, colds, and tumours. Tossa jute (*Corchorus olitorius* L.) is one of the Tunisian plants, traditionally which have great potential on the medicinal purpose. Tossa jute leaves are used in domestic preparation for their nutritive values. They are also employed as a medicine thanks to their diuretic, antipyretic, analgesic, and antimicrobial activities, and their interesting content in antitumor [73] and phenolic antioxidative compounds (**Table 9**; **Figure 19**). Tossa jute is known as Egyptian spinach (*Senaung betina*) and its leaves are rich in potassium, calcium, phosphorus, iron, ascorbic acid, and carotene. There is good source of protein, folate, magnesium, vitamins, and dietary fibre. Jute plant acts as natural antioxidants that prevent and treat age-related diseases. It helps to treat diabetes, obesity, inflammation, pain, fever, ulcer, and heart disease. Leaves of this plant are an excellent source of omega-3 fatty acids (>49 %) than any other vegetable. The leaves help to reduce arsenic and lead toxicity and cholesterol levels as well as infection. The seeds of the two species can be

Sl. No.	Antioxidative compounds	Content (mg/100 g of fresh wt)
1	5-Caffeoylquinic acid	383.9 ± 20.4
2	3,5-Dicaffeoylquinic acid	102.1 ± 8.3
3	Quercetin 3-galactoside	53.3 ± 5.0
4	Quercetin 3-glucoside	376.8 ± 2.8
5	Quercetin 3-(6-malonylglucoside)	126.2 ± 10.4
6	Quercetin 3-(6-malonylgalactoside)	16.7 ± 1.0
7	Ascorbic acid	257.8 ± 14.7
8	R-tocopherol	14.0 ± 0.7

Values are mean (SD of three replications).

Table 9.
Contents of antioxidative compounds in C. olitorius *leaves [74].*

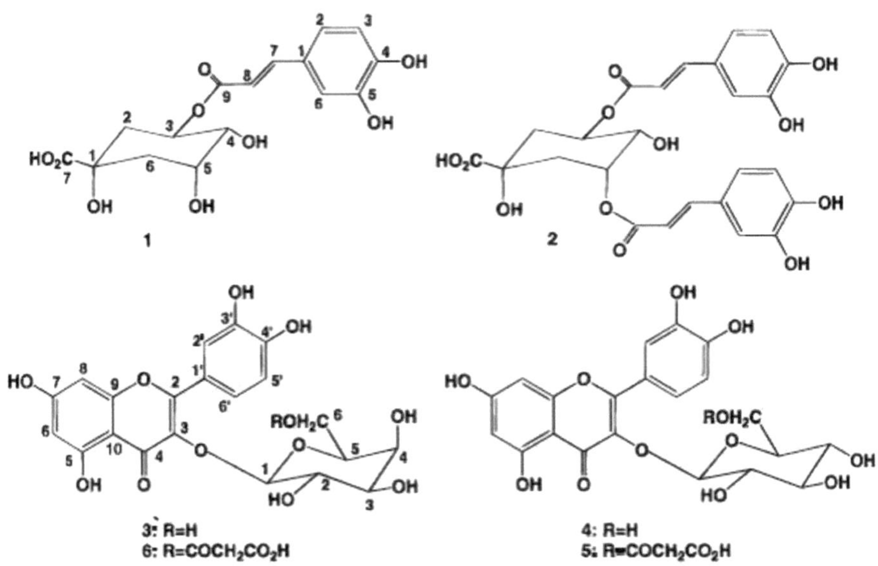

Figure 19.
Chemical structures of phenolic antioxidants isolated from C. olitorius *leaves [74].*

differentiated morphologically; thus, *C. capsularis* are relatively larger in size, irregular in shape, and coppery red when mature, whereas the seeds of *C. olitorius* are somewhat triangular and greyish-green or bluish-black in colour. There are bitter Constituents of the Seeds of *Corchorus olitorius* L., '*Corchorgenin*'-A New Cardiac-active Aglycone [75].

Although kenaf is a fibre crop and used by the fibre industry, the whole immature kenaf plant, stalk (core and bark), and leaves could be considered as high-quality livestock feed. Crude protein in kenaf leaves, stalk crude, and whole plant ranged from 14 to 34%, 2 to 12%, and 6 to 23%, respectively. Kenaf leaves consist of higher quantity of nitrogen compared with stalk, during the growing season. Kenaf has reasonable digestibility with great percentage of digestible protein and hence can be ensilaged efficiently. Dried kenaf leaves are turned into different-sized pellets as high-protein feed source for chickens, rabbits, fish, and goats. Chopped kenaf meal is also used for sheep and Spanish goats as a supplement. Kenaf leaves are edible within 10 days of planting, are rich in protein, that is, about 34%, and are delicious. Kenaf leaf recipes are important edible items in Haiti and used extensively in salads, soups, boiled like spinach, or added to rice, and can also be used for baking. Kenaf seed oils displayed greater antioxidant activity compared with all traditional edible oils ($P < 0.05$) because of the considerable unique composition and oil content of kenaf seed oil. Thus, it becomes an unconventional, cost-effective, and tremendous source of solvent-free vegetable cooking oil with extraordinary antioxidant properties for human consumption [60].

14. Conclusion

Jute and allied fibre crops have a lot of significant uses along with the fibre production. These are natural plants having nutritional, cosmetic, pharmacological uses. The jute and allied fibre crops would be considered as good sources of essential amino acids, phytochemicals having antioxidants activity, antidiabetic activity, hepatoprotective activity, antimicrobial activity, antitumour/anticancer activity, cardioprotective activity, neuroprotective activity, analgestic activity, wound-healing activity, toxicity effects etc. The study of nutritional importance should be emphasized to discover their large-scale utilization. These plant extracts and vegetables would be useful for our health. This will enhance our immunity system against various diseases.

Acknowledgements

The author expressed his gratitude to all scientists and Head of the Breeding Division; researchers & staffs of various levels, and honourable Director General of Bangladesh Jute Research Institute who collaborated and supported him to prepare this chapter through providing the necessary documents and valuable information. He also thanked the Officers of the Department of Agricultural Extension (DAE) and the farmers of Bangladesh who are directly involved with the cultivation of jute and allied fibre crops intentionally for financial income, nutritional purposes as well as making the soil fertile and eco-friend environment. Mohammad Mia Mukul was very pleased to his wife (Tumpa Rahman, Kushtia Govt. College, Bangladesh) who inspired, supported, and technically helped him in writing this informative chapter on nutritional values of JAF crops. He also thanked his honourable parents, teachers, colleagues, friends who always inspired him in writing scientific papers, and chapters on new issues important for the nation.

Abbreviations

BJMC	Bangladesh Jute Mill Corporation
BJRI	Bangladesh Jute Research Institute
CRIJAF	Central Research Institute for jute and Allied Fibres
DAE	Department of Agricultural Extension
DASS	Days after seed sowing

ICDDR'B	International Centre for Diarrhoeal Disease Research, Bangladesh
ICJC	Indian Central Jute Committee
JARL	Jute Agricultural Research Laboratory
USDA	United States Department of Agriculture

Author details

Mohammad Mia Mukul
Breeding Division, Bangladesh Jute Research Institute, Ministry of Agriculture, Dhaka, Bangladesh

*Address all correspondence to: mukulbjribreeding@gmail.com

References

[1] Maity S, Datta AK. Cytomorphological studies in F_1 hybrids (*Corchorus capsularis* L. × *Corchorus trilocularis* L.) of jute (*Tiliaceae*). Comp. *Cytogenet*. 2008;**2**:143-149

[2] Mukul MM, Akter N, Ahmed SSU, et al. Genetic diversity analyses of twelve Tossa jute (*Corchorus olitorius* L.) genotypes based on variability, heritability and genetic advance for yield and yield attributing morphological traits. International Journal of Plant Breeding and Genetics. 2020;**14**:9-16. DOI: 10.3923/ijpbg.2020.9.16

[3] Islam and Rahman, 2008. Hand Book on Agricultural Technologies of Jute, Kenaf and Mesta Crops. Bangladesh Jute Research Institute, Ministry of Agriculture, Manik Mia Avenue, Dhaka-1207. Available from: www.bjri.gov.bd

[4] Bertoia ML, Mukamal KJ, Cahill LE, Hou T, Ludwig DS, Mozaffarian D, et al. Changes in intake of fruits and vegetables and weight change in United States men and women followed for up to 24 years: Analysis from three prospective cohort studies. PLoS Medicine. 2016;**13**(1):e1001956. DOI: 10.1371/journal.pmed.1001956

[5] Choudhary SB, Sharma HK, Karmakar PG, et al. Nutritional profile of cultivated and wild jute (*Corchorus*) species. Australian Journal of Crop Science. 2013;**7**(13):1973-1982

[6] Aune D, Giovannucci E, Boffetta P, Fadnes LT, et al. Fruit and vegetable intake and the risk of cardiovascular disease, total cancer and all-cause mortality—A systematic review and dose-response meta-analysis of prospective studies. International Journal of Epidemiology. 2017;**46**(3): 1029-1056. DOI: 10.1093/ije/dyw319

[7] Amorim A. Population genetics. Brenner's Encyclopedia of Genetics. 2013;**5**:407-411. DOI: 10.1016/b978-0-12-374984-0.01195-5

[8] Burden CJ. Population genetics. Reference Module in Life Sciences. Encyclopedia of Bioinformatics and Computational Biology. 2019;**2**:759-788. DOI: 10.1016/b978-0-12-809633-8.90693-9

[9] Clark AG. Population genetics. Encyclopedia of Genetics. 2001:1513-1519. DOI: 10.1006/rwgn.2001.1016

[10] Ghosh S, Meena K, Sinha MK, Karmakar PG. Genetic diversity in *Corchorus olitorius* genotypes using jute SSRs. Proceedings of the National Academy of Sciences, India Section B: Biological Sciences. 2015;**87**(3):917-926. DOI: 10.1007/s40011-015-0652-4

[11] BJRI. Leaflet: BJRI Tossa Pat 2 (O-9897). Bangladesh Jute Research Institute, Agriculture Wing, Sher-E Bangla Nagar, Dhaka-1207. 2004. Available from: www.bjri.gov.bd

[12] Ma H, Yang R, Song L, Yang Y, Wang Q, Wang Z, et al. Differential proteomic analysis of salt stress response in jute (*Corchorus capsularis & olitorius* L.) seedling roots. Pakistan Journal of Botany. 2015;**47**:385-396

[13] Roy A, Bandyopadhyay A, Mahapatra AK, Ghosh SK, Singh NK, Bansal KC, et al. Evaluation of genetic diversity in jute (*Corchorus* spp.) using STMS, ISSR and RAPD markers. Plant Breeding. 2006;**125**(3):292-297. DOI: 10.1111/j.1439-0523.2006.01208.x

[14] Yang Z, Lu R, Dai Z, Yan A, Chen J, Bai Z, et al. Analysis of genetic diversity

and population structure of a worldwide collection of *Corchorus olitorius* L. germplasm using microsatellite markers. Biotechnology & Biotechnological Equipment. 2018;**32**(4):961-967. DOI: 10.1080/13102818.2018.1438852

[15] Al-Mamun M, Akter N, Saha CK, 2018. BJRI Developed Traditional Varieties of Jute, Kenaf and Mesta. Bangladesh Jute Research Institute, Manik Mia Avenue, Dhaka-1207. Available from: www.bjri.gov.bd

[16] Mukul MM. Elucidation of genotypic variability, character association, and genetic diversity for stem anatomy of twelve Tossa jute (*Corchorus olitorius* L.) genotypes. Hindawi BioMed Research International. 2020;**2020**:16. DOI: 10.1155/2020/9424725

[17] Mukul MM, Akter N. Morpho-anatomical variability, principal component analysis and Euclidean clustering of tossa jute (*Corchorus olitorius* L.). Heliyon, Elsevier. 2021;**7**(5): 1-17. DOI: 10.1016/j.heliyon.2021.e07042

[18] Kundu A, Topdar N, Sarkar D, et al. Origins of white (Corchorus capsularis L.) and dark (C. olitorius L.) jute: A re-evaluation based on nuclear and chloroplast microsatellites. Journal of Plant Biochemistry and Biotechnolog. 2012;**22**:372-381. DOI: 10.1007/s13562-012-0165-7

[19] Zeghichi S, Kallithraka S, Simopoulos AP. Nutritional composition of Molokhia (*Corchorus olitorius*) and Stamnagathi (*Cichorium spinosum*). Plants in Human Health and Nutrition Policy. 2003;**91**:1-21. DOI: 10.1159/000069924

[20] Islam MS, Saito JA, Emdad EM, Ahmed B, Islam MM, Halim A, et al. Comparative genomics of two jute species and insight into fibre biogenesis. Nature Plants. 2017;**3**:16223. DOI: 10.1038/nplants.2016.223

[21] Zhang L, Ma X, Zhang X, Xu Y, Ibrahim AK, et al. Reference genomes of the two cultivated jute species. Plant Biotechnology Journal. 2021;**19**(11): 2235-2248. DOI: 10.1111/pbi.13652

[22] Bhandari HR, Bhanu AN, Srivastava K, et al. Assessment of genetic diversity in crop plants—An overview. Advances in Plants & Agriculture Research. 2017;7 (3):279-286. DOI: 10.15406/apar.2017.07.00255

[23] Jui SA, Mukul MM, Rashid MHO, Nur IJ, Ghosh RK, Mostofa MG, et al. Responses and screening of white jute (*Corchorus capsularis* L.) genotypes against salinity stresses. Plant Science Today, India. 2021;**8**(2):416-424. DOI: 10.14719/pst.2021.8.2.1083

[24] Mukul MM, Akter N, Islam MM, Bhuiyan MSH, Mostofa MG, Ghosh RK, et al. Morpho-phenetical study of high yielding tossa jute variety BJRI Tossa Pat 7 (MG-1) for bast fibre yield and qualities. Heliyon, Elsevier. 2021;**7**(10): 1-17. DOI: 10.1016/j.heliyon.2021.e08129

[25] Mukul MM, Akter N, Mostofa MG, Ahmed SSU, Nur IJ, Al-Mamun M, et al. Analyses of genetic variability, character association, heritability and genetic advance of Tossa jute (*Corchorus olitorius*) genotypes for morphology & stem anatomy. American Journal of BioScience, U.S.A. 2020;**8**(4):99-112. DOI: 10.11648/j.ajbio.20200804.12

[26] Chowdhury GW, Koldewey HJ, Duncan E, Napper IE, Niloy MNH, Nelms SE, et al. Plastic pollution in aquatic systems in Bangladesh: A review of current knowledge. Science of the Total Environment. 2020;**761**:143285. DOI: 10.1016/j.scitotenv.2020.1432

[27] Mariod AA, Saeed Mirghani ME, Hussein I. *Hibiscus cannabinus L.* Kenaf. Unconventional Oilseeds and Oil Sources. 2017:45-51. DOI: 10.1016/b978-0-12-809435-8.00009-3

[28] Mostofa MG, Rahman L, Muhammad Yahiya AS, Harun-or-Rashid M, Mukul Mia M. RAPD analysis of genetic diversity and relationships among Kenaf (*Hibiscus cannabinus* L.) germplasm. Indian Journal of pure and applied BioSciences, India. 2020;**8**(3):37-47. DOI: 10.18782/2582-2845.8096

[29] Nur IJ, Kundu B, Chowdhury MAH, Mukul MM, Ferdush J, Khan MA. Weed Management of Kenaf (*Hibiscus Cannabinus*) through intercropping leafy vegetables and cultural practices. SAARC Journal of Agriculture. 2021;**19**(1):165-176. DOI: 10.3329/sja.v19i1.54787

[30] Islam MM. Jute, Kenaf and Mesta crops research reviews. Letter press Katabon, Dhaka-1205, Bangladesh. 2020. Available from: http://old.bansdoc.gov.bd/iaeabooks/IAEA%20Books/477%20Jute%20Kenaf%20and%20Mesta%20Crops%20Research%20Reviews.pdf

[31] Islam MM. Varietal advances of jute, Kenaf and Mesta crops in Bangladesh: A review. International Journal of Bioorganic Chemistry. 2019;**4**(1):24-41. DOI: 10.11648/j.ijbc.20190401.15

[32] Mukul MM, Akter N, Mostofa MG, Rahman MS, Hossain MA-E, Roy DC, et al. Analyses of variability, Euclidean clustering and principal components for genetic diversity of eight Tossa jute (*Corchorus olitorius* L.) genotypes. Plant Science Today, India. 2020;**7**(4):564-576. DOI: 10.14719/pst.2020.7.4.854

[33] Haque SMA, Rashid MH, Hossain MB, Haque MS.Training handbook on production technology of jute and allied fibre crops. Bangladesh Jute Research Institute, Manik Mia Avenue, Dhaka-1207. 2021. Available from: www.bjri.gov.bd

[34] Mahapatra BS, Mitra S, Ramasubramanian T, Sinha MK. Research on jute (*Corchorus olitorius* and *C. capsularis*) and kenaf (*Hibiscus cannabinus* and *H. sabdariffa*): Present status and future perspective. Indian Journal of Agricultural Science. 2009;**79**(12):951-967

[35] Randhawa MA, Khan AA, Javed MS, Sajid MW. Green leafy vegetables: A health promoting source. Handbook of Fertility. 2015:205-220. DOI: 10.1016/b978-0-12-800872-0.00018-4

[36] Ali MM, Rahman MT, Ahmed T, Rokeya B, Roy B. Advance research on nutraceutical composition of mature jute leaves. International Journal of Recent Innovations in Medicine and Clinical Research. 2020;**2**(4):124-137

[37] Akindahunsi AA, Salawu SO. Phytochemical screening and nutrient-antinutrient composition of selected tropical green leafy vegetables. African Journal of Biotechnology. 2005;**4**(6):497-501

[38] Dairo FA, Adanlawo IG. Nutritional quality of Crassocephalum crepidioides and Senecio biafrae. Pakistan Journal of Nutrition. 2007;**6**(1):35-39. DOI: 10.3923/pjn.2007.35.39

[39] Adepoju OT, Onasanya LO, Udoh CH. Comparative studies of nutrient composition of cocoyam (*Colocassia esculenta*) leaf with some green leafy vegetables. Nigerian Journal of Nutritional Sciences. 2006;**27**(1):22-26

[40] Emebu PK, Anyika JU. Proximate and mineral composition of kale (*Brassica oleracea*) grown in Delta State, Nigeria. Pakistan Journal of Nutrition.

2011;**10**(2):190-194. DOI: 10.3923/pjn.2011.190.194

[41] Akubugwo IE, Obasi NA, Chinyere GC, Ugbogu AE. Nutritional and chemical value of *Amaranthus hybridus* L. leaves from Afikpo, Nigeria. African Journal of Biotechnology;**6**(24): 2833-2839

[42] Mensah JK, Okoli RI, Ohaju-Obodo JO, Eifediyi K. Phytochemical, nutritional and medical properties of some leafy vegetables consumed by Edo people of Nigeria. African Journal of Biotechnology. 2008;**7**(14): 2304-2309

[43] Islam MM. Biochemistry, medicinal and food values of jute (*Corchorus capsularis* L. and *C. olitorius* L.) review. International Journal of Enhanced Research in Science Technology & Engineering. 2013;**2**(11):35-44

[44] Nasir. 2019. Nutritive value of indigenous leafy vegetables of Khulna (MS Thesis). Food and Nutrition Department, KCC Women's College (Affiliated by Khulna University) Khulna, Bangladesh

[45] Mukul MM, Ahmed SSU, Akter N, Mostofa MG, Rahman MS, Talukder FU. Responses of seed germination, seedling growth under salinity stresses and variability for phenotypic traits in Tossa jute (*Corchorus olitorius* L.). Plant Science Today, India. 2020;**8**(2):416-424. DOI: 10.14719/pst.2021.8.1.999

[46] Khatun R, Hossain MA, Parvin MN. Leaflet: BJRI Deshi Pat Shak 1 (BJC-390). Breeding Division, Bangladesh Jute Research Institute (www.bjri.gov.bd), Ministry of Agriculture, Manik Mia Avenue, Dhaka-1207 & College Gate Binding & Printing, Dhaka, Bangladesh; 2015

[47] Akter N, Mostofa MG, Mollah AF, Ghosh RK, Rashid MH. Leaflet: BJRI Deshi Pat Shak 2 (Mehra Red). Agriculture Wing, Bangladesh Jute Research Institute, Manik Mia Avenue, Dhaka-1207. 2020. Available from: www.bjri.gov.bd

[48] Akter N, Mostofa MG, Mollah AF, Ghosh RK, Rashid MH. Leaflet: BJRI Deshi Pat Shak 3 (Mehra Green). Agriculture Wing, Bangladesh Jute Research Institute, Manik Mia Avenue, Dhaka-1207. 2020. Available from: www.bjri.gov.bd

[49] BJRI. Leaflet: BJRI Tossa Pat 3 (OM-1). Bangladesh Jute Research Institute, Agriculture Wing, Sher-E Bangla Nagar, Dhaka-1207. 2005. Available from: www.bjri.gov.bd

[50] ICAR. ICAR-annual report (2018-19). In: Indian Council of Agricultural Research, Central Research Institute for Jute and Allied Fibres (CRIJAF). Barrackpore, Kolkata, West Bengal: India; 2018. Available from: https://crijaf.icar.gov.in/pdf/anrep/an_rep_2018-19.pdf

[51] Sarkar D, Mahato AK, Satya P, Kundu A, Singh S, Jayaswal PK, et al. The draft genome of *Corchorus olitorius* cv. JRO-524 (Navin). Genomics Data. 2017;**12**:151-154. DOI: 10.1016/j.gdata.2017.05.007

[52] Ayadi R, Hanana M, Mzid R, Hamrouni L, Khouja ML, Hanachi AS. *Hibiscus cannabinus* L. – « Kenaf »: A review paper. Journal of Natural Fibres. 2016;**14**(4):1-19. DOI: 10.1080/15440478.2016.1240639

[53] Afzal MZ, Ibrahim AK, Xu Y, Niyitanga S, Li Y, Li D, et al. Kenaf (*Hibiscus cannabinus* L.) breeding. Journal of Natural Fibres. 2020;**17**:1-19. DOI: 10.1080/15440478.2020.1852998

[54] Saba N, Paridah MT, Jawaid M, Abdan K, Ibrahim NA. Potential utilization of Kenaf biomass in different applications. Agricultural Biomass Based Potential Materials. 2015:1-34. DOI: 10.1007/978-3-319-13847-3_1

[55] Islam MR, Rahman MM, Begum HA. Leaflet: Kenaf variety: HC-2 *(Hibiscus cannabinus)*. Bangladesh Jute Research Institute, Manik Mia Avenue, Dhaka-1207. 2000. Available from: www.bjri.gov.bd

[56] Alexopoulou E, Papatheohari Y, Christou M, Monti A. Origin, Description, Importance, and Cultivation Area of Kenaf. In: Monti A, Alexopoulou E, editors. Kenaf: A Multi-Purpose Crop for Several Industrial Applications. London.: Green Energy and Technology. Springer; 2013. DOI: 10.1007/978-1-4471-5067-1_1

[57] FAO (Food and Agricultural Organization) 2003. Consultation on Natural Fibres, The production and consumption of kenaf in China. ESC-Fibers Consultation NO: 03/6. Available from: www.fao.org

[58] Mariod AA, Fathy SF, Ismail M. Preparation and characterisation of protein concentrates from defatted kenaf seed. Food Chemistry. 2010;**123**(3):747-752. DOI: 10.1016/j.foodchem.2010.05.045

[59] Kim DG, Ryu J, Lee MK, et al. Nutritional properties of various tissues from new Kenaf cultivars. Journal of Crop Science and Biotechnology. 2018, 2018;**21**:229-239. DOI: 10.1007/s12892-018-0039-0

[60] Chan KW, Ismail M. Supercritical carbon dioxide fluid extraction of *Hibiscus cannabinus* L. seed oil: A potential solvent-free and high anti-oxidative edible oil. Food Chemistry. 2009;**114**(3):970-975. DOI: 10.1016/j.foodchem.2008.10.055

[61] Ghafar SAA, Yazan LS, Tahir PM, Ismail M. Kenaf seed supercritical fluidextract reduces aberrant crypt foci formation in azoxymethane-induced rats. Experimental and Toxicologic Pathology. 2012;**64**(3):247-251. DOI: 10.1016/j.etp.2010.08.016

[62] Zawawi N, Gangadharan P, Zaini RA, Samsudin MG, Karim R, Maznah I. Nutritional values and cooking quality of defatted kenaf seeds yellow (DKSY) noodles. Inter-national Food Research Journal. 2014;**21**(2):603-608

[63] Cheng WY, Akanda JMH, Nyam KL. Kenaf seed oil: A potential new source of edible oil. Trends in Food Science & Technology. 2016;**52**:57-65. DOI: 10.1016/j.tifs.2016.03.014

[64] Williams D, Verghese M, Walker LT, Boateng J, Shackelford L, Chawan CB. Flaxseed oil and flax seed meal reduce the formation of aberrant crypt foci (ACF) in azoxymethane-induced colon cancer in fisher 344 male rats. Food and Chemical Toxicology. 2007;**45** (1):153-159. DOI: 10.1016/j.fct.2006.08.014

[65] Chan KW, Khong NM, Iqbal S, Mansor SM, Ismail M. Defatted kenaf seed meal (DKSM): Prospective edible flour from agricultural waste with high antioxidant activity. LWT–Food Science and Technology. 2013;**53**(1):308-313. DOI: 10.1016/j.lwt.2013.01.003

[66] Jin CW, Ghimeray AK, Wang L, Xu ML, Piao JP, Cho DH. Far infrared assisted kenaf leaf tea preparation and its effect on phenolic compounds, antioxidant and ACE inhibitory activity. Journal of Medical Plants Research. 2013; 7(17):1121-1128. DOI: 10.5897/JMPR 11.1431

[67] Mohamed A, Bhardwaj H, Hamama A, Webber C. Chemical composition of Kenaf (*Hibiscus cannabinus* L.) seed oil. Industrial Crops and Products. 1995; **4**(3):157-165. DOI: 10.1016/0926-6690(95)00027-a

[68] Crane JC. Roselle—A potentially important plant fibre. Economic Botany. 1949;**3**:89-103. DOI: 10.1007/BF02859509

[69] Riaz G, Chopra R. A review on phytochemistry and therapeutic uses of *Hibiscus sabdariffa* L. Biomedicine & Pharmacotherapy. 2018;**102**:575-586. DOI: 10.1016/j.biopha.2018.03.023

[70] Ali BH, Wabel NA, Blunden G. Phytochemical, pharmacological and toxicological aspects of *Hibiscus sabdariffa* L.: A review. Phytotherapy Research. 2005;**19**(5):369-375. DOI: 10.1002/ptr.1628

[71] Rao PU. Nutrient composition and biological evaluation of Mesta (*Hibiscus sabdariffa*) seeds. Plant Foods for Human Nutrition. 1996;**49**(1):27-34. DOI: 10.1007/BF01092519

[72] Babalola SO, Babalola AO, Aworth OC. Compositional attributes of the calyces of Roselle. Journal of Food Technology in Africa. 2001;**6**(4): 133-134. DOI: 10.4314/jfta.v6i4.19306

[73] Zakaria ZA, Somchit MN, Zaiton H, Mat Jais AM, Sulaiman MR, Farah WO, et al. The in vitro antibacterial activity of *Corchorus olitorius* extracts. International Journal of Pharmacology. 2006;**2**: 213-215. DOI: 10.3923/ijp.2006.213.215

[74] Azuma K, Nakayama M, Koshioka M, Ippoushi K, Yamaguchi Y, Kohata K, et al. Phenolic antioxidants from the leaves of *Corchorus olitorius* L. Journal of Agricultural and Food Chemistry. 1999; **47**(10):3963-3966. DOI: 10.1021/jf990347p

[75] Chakrabarti JK, Sen NK. Bitter constituents of the seeds of *Corchorus olitorius* L., "Corchorgenin"—A new cardiac-active aglycone. Journal of the American Chemical Society. 1954;**76**(9): 2390-2392. DOI: 10.1021/ja01638a028